TECHNICIAN'S GUIDE TO FIBER OPTICS
SECOND EDITION

TECHNICIAN'S GUIDE TO FIBER OPTICS

SECOND EDITION

Donald J. Sterling, Jr.

Delmar Publishers Inc.

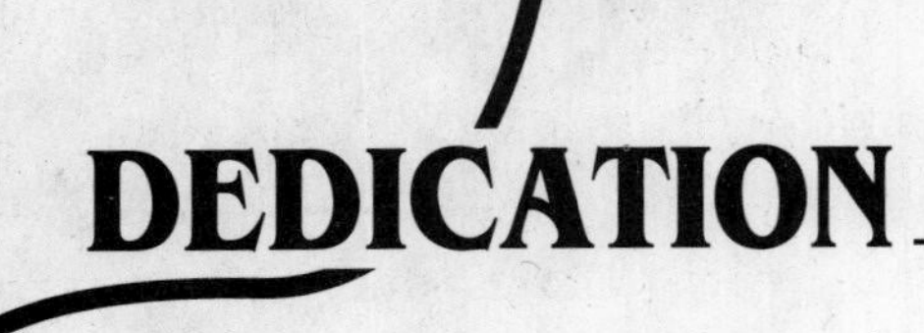

DEDICATION

To Lynne and to Megan

Delmar Staff
Executive Editor: Michael McDermott
Administrative Editor: Wendy J. Welch
Project Editor: Carol Micheli
Production Coordinator: James Zayicek
Art Coordinator: Brian Yacur
Design Coordinator: Lisa Pauly
Cover Photo: Courtesy of AMP Incorporated
Cover Design: Juanita Brown

For information, address Delmar Publishers Inc.
3 Columbia Circle, Box 15–015
Albany, New York 12212-5015

Printed in the United States of America
Published simultaneously in Canada
by Nelson Canada,
A division of The Thomson Corporation

1 2 3 4 5 6 7 8 9 10 XXX 99 98 97 96 95 94 93

Library of Congress Cataloging-in-Publication Data

Sterling, Donald J., 1951–
Technicians guide to fiber optics / Donald J. Sterling, Jr.
p. cm.
Includes index.
ISBN 0-8273-5835-0
1. Fiber optics. I. Title.
TA1800.S74 1993
621.36'92—dc20 92-32948
CIP

CONTENTS

PREFACE

The first edition of *Technician's Guide to Fiber Optics* was well received as an introductory general text on fiber optics. It is suitable for technicians and students in technical and trade schools, community colleges, and first- or second-year students at four-year colleges. It also is well suited for use by companies who wish to acquaint employees with this new technology.

This new edition follows the first edition very closely. It has been updated to include such advances in technology as erbium-doped fiber amplifiers. But the greatest change in fiber optics since the first edition can be summed up in one word: *standardization.* Chapter 11 has been heavily revised and Chapter 15 has been completely rewritten to emphasize standard components and applications. To this end, FDDI is extensively covered, as are optical LANs and telecommunications.

The book is intended to be self-contained. Few prerequisites are required, although readers should be familiar with basic electronics and digital concepts. These topics are briefly reviewed but not developed much past the level needed to understand them within the context of fiber optics, as discussed here.

Use of mathematics is kept limited to practical applications most useful to students. Only a basic acquaintance with trigonometry and logarithms is required.

The book is divided into three parts. Part One, Chapters 1 through 4, attempts to put fiber optics into perspective as a transmission medium. It not only describes the advantages of this new technology over copper counterparts, but also it shows the growing importance of electronic communications and explains basic concepts such as bits and bytes, analog and digital, and light.

Part Two, Chapters 5 through 12, describes in detail fibers, sources, detectors, connectors and splices, and couplers. Emphasis is given to the theoretical and practical aspects of fibers and connectors in particular, because these materials are the most different from their electronic counterparts. Understanding these components will be of most value to anyone actually working with fiber optics.

Part Three, Chapters 13 through 16, attempts to show how fiber-optic systems are put together. It includes discussion of links (with emphasis on how to calculate power and rise-time budgets), installation and special fiber-optic hardware, applications, and fiber-optic equipment.

Each chapter contains a summary of important concepts and a set of review questions. A helpful glossary of key terms completes the book.

Again I would like to thank all those who helped make this a better book: Denise Forster of BT&D, Steve Dawson of Tektronix, Susan Ursch of Digital Equipment Corporation, Mitch Strobin of Network Peripherals were all kind enough to provide photographs for this work.

While many people at AMP Incorporated offered their time and expertise, I owe special debts of thanks to Christine Arnold, Marilyn Arnold, Chris Shroyer, Bret Matz, Mike Baum, Bob Southard, Mike Peppler, Kevin Monroe, Jack Himes, Tom Ball, Dennis Hess, Raj Kapany, Terry Bowen and Dave Hubbard.

ACKNOWLEDGMENTS

First, thanks are due to those who reviewed the manuscript and made valuable suggestions for improvements: Howard Griffin, Jerry Harshman, Harbans B. Mathur, Lee Rosenthal, and Jerry L. Wescott. For their indespensable assistance with the first edition, I also thank Elias A. Awad, Carl Stancil, Peter Vangel, Sam Conn, Tom Gerwattowski, and Bruce Bothwell.

I would also like to thank those companies who generously supplied information, artwork, and samples for the book. In particular, I would like to acknowledge the help provided by Loretta Coltrane of GTE, Elaine Hicks of Siecor, Nancy Suey and John Sicotte of Corning, Melissa A. Heck of Anritsu America, Ellen Phillips of Lytel, Carl Blesch of AT&T Bell Laboratories, Marguerite G. Shapalis of ITT Electro-Optical Products, Richard Ryan of Buehler, Andrew Guraxani and A. F. Nelson of Hewlett-Packard, and Robert Chow of Fujitsu Microelectronics.

I would like to thank Richard Aure of Canoga Data Systems for permission to quote from Canoga's application note *Fiberoptic Cable Installation Techniques,* which forms a central part of Chapter 14.

AMP was also kind enough to provide many of the illustrations used in the book as well as permission to quote liberally from material I wrote while in its employ, especially *Designer's Guide to Fiber Optics.*

TRADEMARKS

Efforts have been made to recognize trademarks for products mentioned in this book. The following is a list of trademark holders and their trademarks. We apologize for any missed trademarks.

AMP Incorporated: AMP, DUALAN, QUADLAN; AT&T: ST; International Business Machines Corporation: ESCON, IBM; E.I. du Pont de Nemours & Co., Inc.: Kevlar, Mylar, Teflon; Siecor Corporation: Mini-Bundle and Siecor; NTT Advanced Technology Corporation: SC, FC

Part One

BACKGROUND

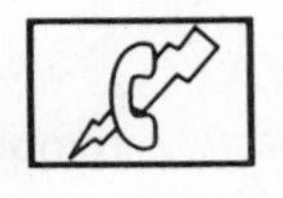

CHAPTER 1 The Communications Revolution

Fiber optics, as discussed in this book, is simply a method of carrying information from one point to another. An optical fiber is a thin strand of glass or plastic that serves as the transmission medium over which the information passes. It thus fills the same basic function as a copper wire carrying a telephone conversation or computer data. But, unlike the copper wire, the fiber carries light instead of electricity. In doing so, it offers many distinct advantages that make it the transmission medium of choice in applications ranging from telephony to computers to automated factories.

The basic fiber-optic system is a link connecting two electronic circuits. Figure 1–1 shows the main parts of a link, which are as follows:

- *Transmitter,* which converts an electrical signal into a light signal. The source, either a light-emitting diode or laser diode, does the actual conversion. The drive circuit changes the electrical signal fed to the transmitter into a form required by the source.
- *Fiber-optic cable,* which is the medium for carrying the light. The cable includes the fiber and its protective covering.
- *Receiver,* which accepts the light and converts it back into an electrical signal. The two basic parts of the receiver are the detector, which converts the light signal back into an electrical signal, and the output circuit, which amplifies and, if necessary, reshapes the electrical signal before passing it on.
- *Connectors,* which connect the fibers to the source, detector, and other fibers.

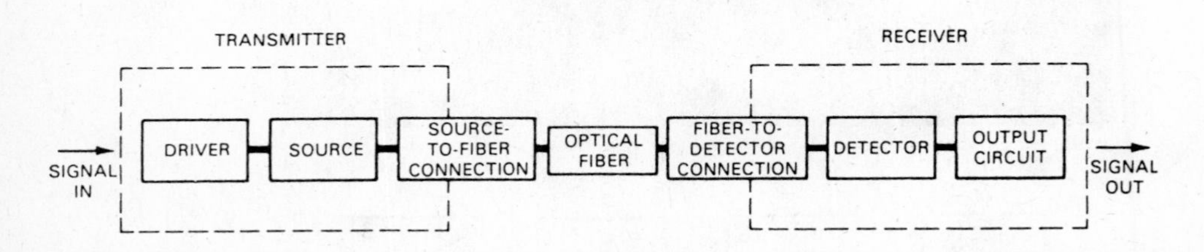

FIGURE 1–1 Basic fiber-optic link (Illustration courtesy of AMP Incorporated)

As with most electronic systems, the transmitter and receiver circuits can be very simple or very complex. These four parts form the essence of the fiber-optic communication alternative. Other components discussed in this book, such as couplers, multiplexers, and distribution hardware, provide the means of forming more complex links and communication networks. But the transmitter, fiber, receiver, and connection are the basic elements found in every fiber-optic link.

THE HISTORY OF FIBER OPTICS

Using light for communication is not new. Lanterns in Boston's Old North Church sent Paul Revere on his famous ride. Navy signalmen used lamps to communicate by Morse code, and lighthouses for centuries have warned sailors of dangers.

Claude Chappe built an optical telegraph in France during the 1790s. Signalmen in a series of towers stretching from Paris to Lille, a distance of 230 km, relayed signals to one another through movable signal arms. Messages could travel end to end in about 15 min. In the United States, an optical telegraph linked Boston and the nearby island of Martha's Vineyard. These systems were eventually replaced by electrical telegraphs.

The English natural philosopher John Tyndall, in 1870, demonstrated the principle of guiding light through internal reflections. In a demonstration before the Royal Society, he showed that light could be bent around a corner as it traveled in a jet of pouring water. Water flowed through a horizontal spout near the bottom of a container into another container and through a parabolic path through the air. When Tyndall aimed a beam of light out through the spout along with the water, his audience saw the light follow a zigzag path inside the curved path of the water. A similar zigzagging occurs in optical fibers.

A decade later, Alexander Graham Bell patented the photophone (Figure 1–2), which used unguided light to carry speech. A series of lenses and mirrors threw

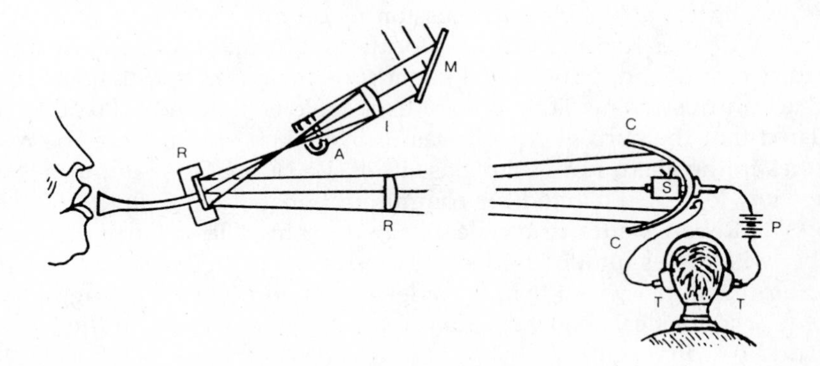

FIGURE 1–2 Alexander Graham Bell's photophone (Illustration courtesy of AMP Incorporated)

light onto a flat mirror attached to the mouthpiece. The voice vibrated the mirror, thereby modulating the light striking it. The receiver used a selenium detector whose resistance varied with the intensity of light striking it. The voice-modulated sunlight striking the selenium varied the current through the receiver and reproduced the voice. Bell managed to transmit over 200 m.

Throughout the early 20th century, scientists made experimental and theoretical investigations into dielectric waveguides, including flexible glass rods.

During the 1950s, image-transmitting fibers were developed by Brian O'Brien at the American Optical Company and by Narinder S. Kapany and colleagues at the Imperial College of Science and Technology in London. Such fibers find use in fiberscopes, which are used in medicine to look inside the body. It was Kapany who invented the glass-coated glass rod and coined the term *fiber optics* in 1956. In 1973, Dr. Kapany founded Kaptron, a company specializing in fiber-optic couplers and switches as discussed in Chapter 12.

In 1957, Gordon Gould, a graduate student at Columbia University, described the laser as an intense source of light. Charles Townes and Arthur Schawlow of Bell Laboratories helped popularize the idea of a laser in scientific circles and helped spur research into devising an operating laser. By 1960, Theodore Maiman of Hughes Laboratories operated the first ruby laser, while Townes demonstrated a helium neon laser. By 1962, lasing was observed in a semiconductor chip, which is the type of laser used in fiber optics. Gould, rather belatedly, was awarded four basic patents for lasers in 1988, based on his work in the 1950s.

The importance of the laser as a carrier of information was not lost on communication engineers. A laser carrier had an information-carrying capacity 10,000 times the magnitude of the radio frequencies then being used. Despite this, the laser was not well suited for open-air line-of-sight transmission. Fog, smog, rain, and other environmental conditions adversely affect the transmission of laser light. It is easier to transmit a laser beam from the earth to the moon than from uptown to downtown Manhattan. The laser, then, was a communications source waiting for a suitable transmission medium.

In 1966, Charles Kao and Charles Hockham, of Standard Telecommunication Laboratory in England, published a paper proposing that optical fibers could be used as a transmission medium if their losses could be reduced to 20 dB/km. They speculated that the current high losses of over 1000 dB/km were the result of impurities in the glass, not of the glass itself. Reducing these impurities would produce low-loss fibers suited for communications.

In 1970, Robert Maurer and colleagues at Corning Glass Works produced the first fiber with losses under 20 dB/km. By 1972, losses were reduced to 4 dB/km in laboratory samples, well below the level Kao and Hockham suggested was required for a practical communication system. Today, losses in the best fibers are around 0.2 dB/km.

Similar advances in semiconductor sources and detectors, connectors, transmission technology, communication theory, and other areas, as well as intense interest in exploiting the possible benefits of fiber optics, led to investigation and trials of fiber-optic systems during the mid- and late 1970s.

The Navy installed a fiber-optic telephone link aboard the USS *Little Rock* in 1973. The Air Force replaced the wiring harness of an A-7 aircraft with an optical link in its Airborne Light Optical Fiber Technology (ALOFT) program in 1976. The wiring harness contained 302 cables with a total length of 1260 m and a weight of 40 kg. The optical replacement had 12 fibers, a total length of 76 m and a weight of 1.7 kg. The military was also responsible for one of the earliest operational fiber-optic links in 1977, a 2-km, 20-Mbps system connecting a satellite earth station and a data-processing center.

In 1977, AT&T and GTE installed fiber-optic telephone systems carrying commercial telephone traffic. Encouragingly, the systems exceeded the rigid performance standards for reliability imposed by the telephone companies on their equipment and its operation. As a result of successful early trials, optical systems were installed by telecommunications companies throughout the late 1970s and early 1980s. In 1980, AT&T announced plans for an ambitious project for constructing a fiber-optic system running from Boston to Richmond. This system demonstrated the acceptance of the technology for normal, high-capacity systems rather than specialized applications. It showed that fiber optics would be the preferred technology in the future. Fiber optics had clearly arrived as a viable technology.

During the same period, technology also advanced rapidly as the industry matured and gained experience. As late as 1983, advanced single-mode fibers (which will be described in Chapter 4) were felt to be so difficult to use that they would find only specialized application for many years. By 1985, major long-distance carriers such as AT&T and MCI had not only installed single-mode systems but also had chosen them as the standard fiber for future installations.

Although industries such as computers, information networks, and industrial controls did not embrace fiber optics as quickly as the military and the telecommunications industry, they nonetheless carefully investigated the technology, experimented with it, and closely observed the experiences of other users. As the information age spread the need for better telecommunication systems, it also fostered the growth of fiber-optic applications. Today, fiber optics finds wide use outside of telecommunications.

For example, IBM, the world's largest computer maker, announced a new mainframe computer in 1990. The channel controller, which links the computer to such peripherals as disk and tape drives, uses fiber optics. This is the first time fiber was offered as *standard* equipment. The benefit of the fiber-based ESCON controller is that it allows faster transfer of information over longer distances. The older copper-based controller had an average transfer rate of 4.5 Mbyte/s over a maximum distance of 400 feet. The new controller offers a transfer rate of 10 Mbyte/s over distances of several miles.

Also in 1990, Linn Mollenauer, a researcher at Bellcore, transmitted a 2.5 Gbps signal over 7500 km *without* regeneration. Normally, a fiber-optic signal must be amplified and returned to its original shape periodically—say every 25 km. As it travels, it loses power and becomes distorted. Mollenauer's system used a soliton laser and erbium-doped, self-amplifying fiber. Soliton pulses do not spread out and lose their wavelike shape as they travel through the fiber. During the same time, Nippon Telephone and Telegraph in Japan achieved 20 Gbps speeds over much shorter distances. The importance of soliton technology is that it will allow a fiber-optic telephone system to span the Pacific or Atlantic without the need for electronic amplifiers along the way. As of 1992, however, soliton technology remains in the lab and does not yet find commercial application. Solitons and erbium-doped fiber amplifiers are further discussed in Chapter 15.

THE INFORMATION AGE

America, it has been said, is entering the information age, which also means the electronic age. The information age uses electronics to perform the four functions of information processing.

1. Gathering information
2. Storing information
3. Manipulating and analyzing information
4. Moving information from one place to another

The tools of the age are fairly new: computers, electronic offices, a complex telephone network, satellites, information networks and databases, cable television, and so forth. If you look, the evidence of the age surrounds you. The information-services industry has been growing at a rate of 15% a year and shows no signs of slowing down.

In short, America—and much of the rest of the world—is quickly becoming a huge network wired together by electronics. As easily as you can telephone from New York to San Francisco, you can send computer data across the country. In offices, large and small computers and other electronic equipment are connected into networks. Engineers use computers to work to design new computers. Home and personal computers abound. This book was written on a computer. It is even possible to write, produce, and publish a book without ever touching a piece of paper. All the information in the book is moved electronically from the writer's word processor to the printer's typesetting and printing equipment.

The following facts give some perspective to the importance of electronics in modern life.

- In 1988, there were an estimated 165 million telephones in the United States, compared to only 39 million in 1950. In addition, the services provided by your telephone company are much more sophisticated.

- From 1950 to 1981, the number of miles of wire in the telephone system increased from 147 million to 1.1 billion.
- In 1990, there were an estimated 5 million miles of fiber in the U.S. telephone system. This will increase to 15 million by the year 2000. Each fiber has the capacity to replace several copper cables.
- In 1989, an estimated 10 million personal computers were sold to businesses and consumers in the U.S. The personal computer did not exist in 1976; today it is commonplace and essential office and engineering equipment.
- More than 1500 computer databases are available from which you can obtain information, using a personal computer and the public telephone network.
- FAXes are replacing letters in many business situations.
- The first fiber-optic telephone system, installed in 1977, operated at 44.7 Mbps and was capable of carrying 672 voices simultaneously. Today, Sonet, which is a standardized system for optical telephony, offers a top speed of 10 Gbps—over 200 times faster than the first optical system. Faster speeds are planned but are not yet approved and standardized because electronic components are not readily available.

The point is that all these represent sources of information and the means of tying them together. The information may be a telephone call to a friend or the design of a new widget. Important to the growing amount of information is the means of moving it from one place to another. This movement may be a telephone call across the country or from one office to another office a couple of doors away. Telephone companies increasingly use the same digital techniques to send your voice and computer data. Surprisingly, your voice becomes more like computer data: It is converted into digital pulses—numbers—just like computer data. By making this conversion, the telephone company transmits your telephone call with greater fidelity and reliability. Digital transmission systems—rather than traditional analog systems—are used in most new telephone systems. One study estimates that 34% of telephone central offices used digital transmission equipment in 1984. By 1994, this percentage will increase to 82%. Fiber optics is exceptionally suited to the requirements of digital telecommunications. Coupled with the need to move information is the need to move it faster, more efficiently, more reliably, and less expensively. Fiber optics fills these needs.

THE WIRING OF AMERICA

The oldest and most obvious example of the wiring of the United States is, of course, the telephone network. At one time, the network carried voices almost exclusively, whereas it now carries voice, computer data, electronic mail, and video pictures.

Cable television is another example. It offers the possibility of interactive television. The cable will not only bring you television, but it will allow you to respond to opinion polls, shop at home, make restaurant and airline reservations, and do many other things. In short, interactive cable TV will become two-way communication. Rather than simply provide television broadcasts to your home, it will allow you to send information from your home. Such systems have been tested experimentally in the United States, Europe, and Japan, and some experiments used fiber optics. A fiber-optic cable television system announced in 1991 to be built in Queens, New York, will be capable of providing over 150 channels to a household.

Home banking and financial services, offered in some cities by major banks, are also made possible by the wiring of America. From a personal computer or terminal in your home, you can check account balances, transfer funds between accounts, pay bills, buy stocks and bonds, add to money market accounts, and so forth.

Another area of wiring where computers have invaded is the office. Word processors have replaced secretaries' typewriters. Computers have replaced the accountant's balance books, the manager's spreadsheet, and the engineer's drafting table. Furthermore, a need exists to allow electronic devices to communicate with each other and with the world outside the office. Local area networks (LANs) fill this need by allowing electronic equipment to be tied together electronically. In addition, many networks also connect to the public telephone network. Again, fiber optics has a role to play in the modern office by providing an efficient transmission medium for networked equipment.

Factories, too, now use electronics extensively. Computers control and monitor activities, and robots perform monotonous, dangerous, or tedious tasks. In advanced factories, various activities are tied together electronically so that all operate harmoniously. But the electrical noise caused by the high-powered machinery can interfere with the control signals. Switching of electrical loads and operating equipment such as arc welders or drills produces electrical noise transients. Because of this noise, a great deal of care, such as shielding cables or separating them from power line sources of noise, must be taken to protect the signals on control lines. Because fibers are immune to electrical noise, fiber optics is a good candidate for the factory.

TELECOMMUNICATIONS AND THE COMPUTER

Until a few years ago, a clear distinction existed between what was part of the telephone system and what was a computer system. Telephone companies, for example, were legally forbidden to compete in the computer marketplace. Today, the distinction, if it still exists, has blurred considerably. Computers send data over telephone lines. Telephone systems turn your voice into computerlike data before transmitting it. Telephone and computer companies now often compete in the same market.

The changes causing this blurring are understandable. Advances in electronic technology meant that each industry used each other's technology. After the breakup of AT&T, the world's largest corporation, in 1982, the distinction between telephony and computers dissolved further. The importance to fiber optics is this: The entire electronic network is becoming a single system. It becomes increasingly difficult to say that this part of the system is the telephone company's responsibility, this part is the computer company's, and this part is the building owner's.

One measure of advances brought by the information age, the wiring of America, and the melding of telephone and computers is the services provided by telephone companies. The service most of us have or had—the ability to dial and receive calls—is called POTS (plain old telephone service). Today, many new services, such as call forwarding and automatic callback, are the beginning of what some call PANS (pretty amazing new services). The telephone companies are aiming toward an integrated-services digital network (ISDN) in which voice, video, and data can be exchanged on the telephone network. Anybody can exchange anything with anybody, anywhere, anytime.

THE FIBER-OPTIC ALTERNATIVE

A wired network such as we have broadly discussed in this chapter needs an efficient medium over which to send information. Traditional technology, such as copper cable and microwave transmission, has limitations that are overcome by fiber optics. Copper cables, for example, have limited capacity and are affected by noise. Microwave transmission, although it has very high capacity, is expensive because of the cost of transmission stations and is limited to line-of-sight transmission, which places severe restrictions on possible locations of transmission equipment. Fiber optics offers capacities well beyond those of copper cables and without the high cost and restrictions of microwave facilities. It will even challenge satellites in some applications. Fiber optics is a new technology whose capabilities are only beginning to be realized. Even though fiber optics possesses many advantages over copper cable, remember that fiber optics is in its infancy, whereas copper cable is a mature technology with much less potential for improvement. Fiber optics promises to be an integral part of the information revolution and the networking of the world.

Fiber optics will affect everyone's life whether it is noticed or not. Here are some examples of areas where fiber optics may touch unnoticed on your life: carrying your voice across the country, bringing television to your home by cable, connecting electronic equipment in your office with equipment in other offices, connecting electronic subsystems in your car, monitoring operations in a factory.

Figure 1–3 shows the applications and functions of fiber-optic communication as a tree. Branched in the figure are the main application areas, showing that leaves are part of a great, billowing tree. You can clearly see that fiber optics has

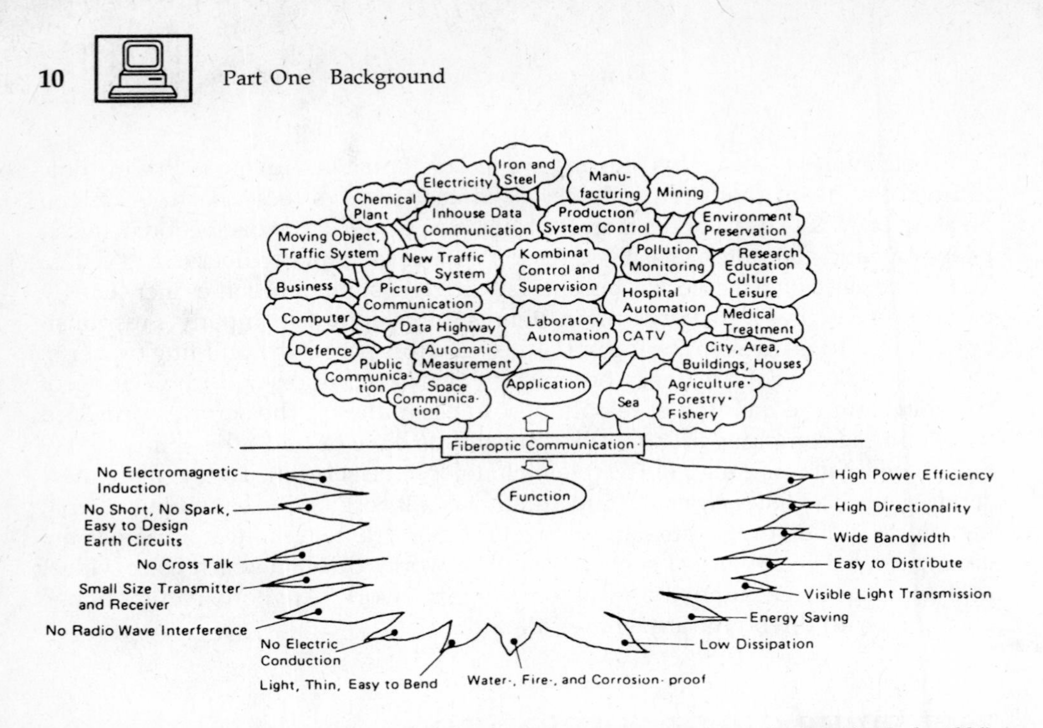

FIGURE 1–3 Applications of fiber optics (Illustration courtesy of Hitachi America Ltd., 1984)

uses in many areas. The functions, or reasons for using fiber optics, are the roots feeding the applications.

Although fiber optics is a new technology, it has a bright future. It has proven itself in many trials and become the transmission medium of choice in many applications. Its advantages also make it a medium of choice in many other applications in the future.

SUMMARY

- The four parts of a fiber-optic link are the transmitter, cable, receiver, and connectors.
- Modern electronics is an important part of the information age.
- Using electronics to gather, store, manipulate, and move information has resulted in the networking of this country.
- The practical use of fiber optics for communication began in the mid- and late 1970s with test trials. Today it is an established technology.
- The networking will continue.
- Fiber optics has an important role to play in this networking and the information age.
- The advantages fiber optics offers over traditional, older transmission media make it attractive in the information age.

? REVIEW QUESTIONS

1. A fiber-optic system transmits information as
   A. Electricity
   B. Light
   C. Sound
   D. Images
2. Name the four main parts of a fiber-optic link.
3. List three of the four functions of information processing.
4. In which of the four areas of information processing is fiber optics most applicable?
5. (True/False) Fiber optics is so specialized a technology that it will never touch most people's lives directly or indirectly.
6. List five areas where fiber optics can be applied. Be as specific as possible.
7. In the 1970s, the greatest interest in and application of fiber optics came from which two of the following:
   A. Military/government
   B. Computer industry
   C. Avionics industry
   D. Telephone industry
   E. Automotive industry
   F. Robotics industry
   G. Medical industry
   H. Power industry
8. (True/False) Light as a communication method awaited the invention of the laser and the optical fiber.

CHAPTER 2

Information Transmission

The last chapter discussed the importance of electronic information to our modern world. The purpose of this chapter is to introduce important aspects of signals and their transmission. An understanding of the underlying principles that allow modern electronic communication is fundamental to an understanding and appreciation of fiber optics. The ideas presented here are fundamental not only to fiber optics but to all electronic communication. We do not propose to offer an in-depth survey of communication theory and its application to electronics. The purpose of this chapter is to introduce terms and principles without which any discussion of fiber optics would be stymied.

COMMUNICATION

Communication is the process of establishing a link between two points and passing information between them. The information is transmitted in the form of a signal. In electronics, a signal can be anything from the pulses running through a digital computer to the modulated radio waves of an FM radio broadcast. Such passing of information involves three activities: encoding, transmission, and decoding.

Encoding is the process of placing the information on a carrier. The vibration of your vocal chords places the code of your voice on air. The air is the carrier, changed to carry information by your vocal chords. Until it is changed in some way, the carrier carries no information. It is not yet a signal. A steadily oscillating electronic frequency can be transmitted from one point to another, but it also carries no intelligence unless information is encoded on it in some way. It is merely the carrier onto which the information is placed. Conveying information, then, is the act of modifying the carrier. This modification is called *modulation*.

Figure 2–1 shows the creation of a signal by impressing information on a carrier. A high-frequency carrier, which in itself carries no information, has impressed on it a lower-frequency signal. The shape of the carrier is now modulated by the information. Although the simple example in the figure conveys very little information, the concept can be extended to convey a great deal of information. A Morse-code system can be based on the example shown. On the

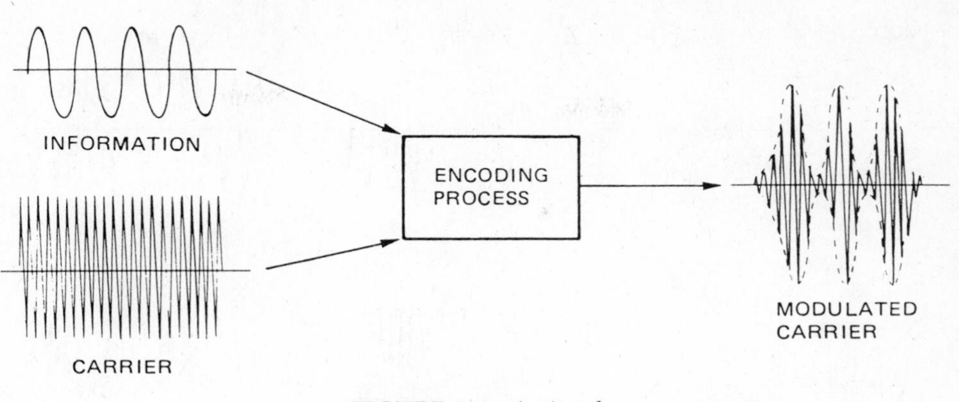

FIGURE 2–1 A signal

unmodulated carrier can be impressed a low frequency for one of two durations corresponding to the dots and dashes of the code.

Once information has been encoded by modulating a carrier, it is *transmitted.* Transmission can be over air, on copper cables, through space to a satellite and back, or through optical fibers.

At the other end of the transmission, the receiver separates the information from the carrier in the *decoding* process. A person's ear separates the vibrations of the air and turns them into nerve signals. A radio receiver separates the information from the carrier. For the encoded signal in Figure 2–1, the receiver would strip away and discard the high-frequency carrier while keeping the low-frequency signal for further processing. In fiber optics, light is the carrier on which information is impressed.

Assume you are writing a letter on a personal computer to a friend. The letters you press at the keyboard are encoded into the digital pulses the computer requires. Once finished, you wish to send the letter to your friend's computer. You will need a *modem,* a device that re-encodes the digital pulses into audio pulses easily sent over the telephone lines to the central telephone office.

The central office receives your transmission. It reencodes your signal into digital pulses and transmits it with many other signals over an optical fiber to another central office. This office decodes the digital pulses to audio pulses. It sends this information to your friend, whose modem further decodes the signal into computer pulses. Your friend can read your letter by displaying it on the computer screen or by printing it out.

Many ways exist to modulate the carrier. Figure 2–2 shows three common ways.

1. *Amplitude modulation (AM).* Amplitude modulation is used in AM radio. Here the amplitude of the carrier wave is varied to correspond to the amplitude of the information.

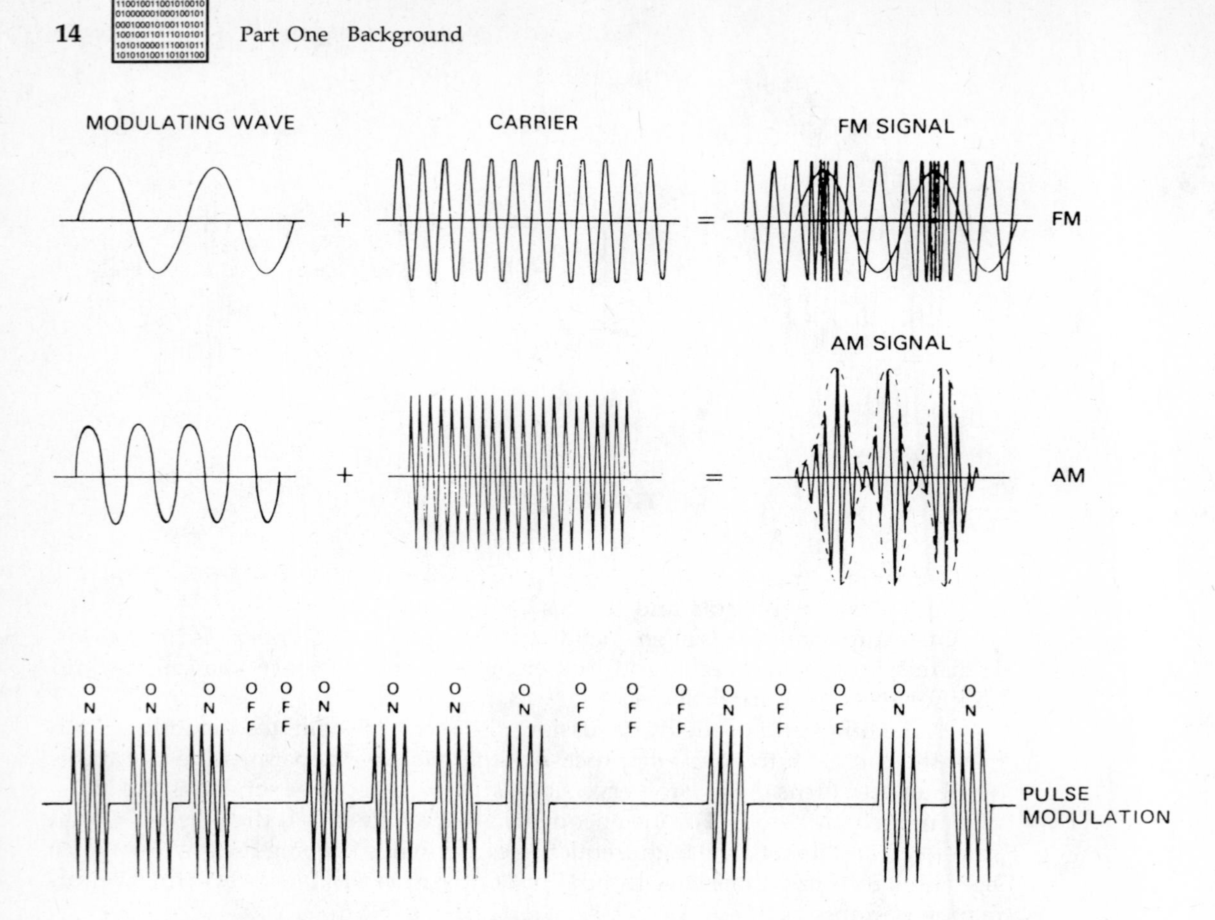

FIGURE 2–2 Examples of modulation (Illustration courtesy of AMP Incorporated)

2. *Frequency modulation (FM).* Frequency modulation changes the frequency of the carrier to correspond to differences in the signal amplitude. The signal changes the frequency of the carrier, rather than its amplitude. FM radio uses this technique.
3. *Pulse-coded modulation (PCM).* Pulse-coded modulation converts an analog signal, such as your voice, into digital pulses. Your voice can be represented as a series of numbers, the value of each number corresponding to the amplitude of your voice at a given instance. PCM is the main way that voices are sent over a fiber-optic telephone system. We will look more closely at it in a moment.

ANALOG AND DIGITAL

We live in an analog world. *Analog* implies continuous variation, like the sweep of the second hand around the face of an electric watch. Your voice is analog

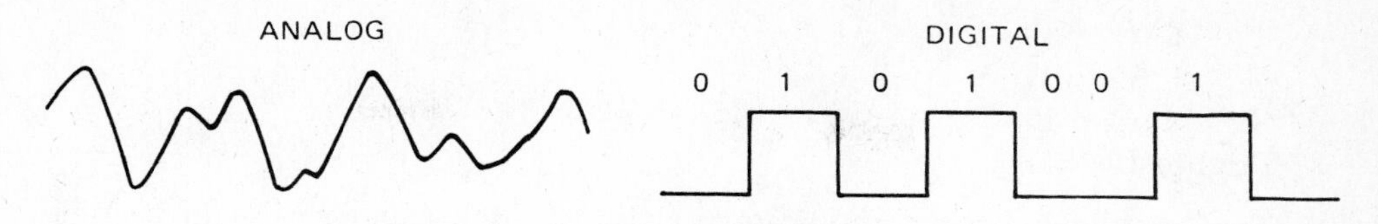

FIGURE 2–3 Analog and digital information

in that the vibrations that form sound can vary to any point within your range of vocal frequencies. Sound is analog, in that it is continuously varying within a given range. Electronic circuits, such as transmitters and receivers, use analog circuits to deal with these continuous variations. Indeed, before the advent of computers, most electronic engineering dealt with analog circuits.

Consider a light operated by a dimmer. Turning the dimmer to adjust the light's brightness is an analog operation. The brightness varies continuously. There are no discrete levels, so that you can easily turn the light more or less bright.

In contrast, *digital* implies numbers—distinct units like the display of a digital watch. In a digital system, all information ultimately exists in numerical values of digital pulses.

In contrast to the dimmer light, a three-way lamp is digital. Each setting brings a specific level of brightness. *No levels exist in between.* Figure 2–3 shows analog and digital signals.

Important to electronic communication is that analog information can be converted to digital information, and digital information can be turned into analog information. You may be familiar with digital stereo systems. In these systems, music is encoded into digital form as a series of numbers that represent the analog variations in the music. The electronics in the playback system must reconvert the digital code into analog music.

Digital Basics: Bits and Bytes

The basis of any digital system is the *bit* (short for *b*inary dig*it*). The bit is the fundamental unit of digital information, and it has only one of two values: 1 or 0. Many ways exist to represent a bit. In electronics, the presence or absence of a voltage level is most common: One voltage level means a *1*; a second level means a *0*. Unfortunately, the single bit 1 or 0 can represent only a single state, such as *on* or *off*. For example, a lamp can be represented by 0 when turned off and by 1 when turned on:

Off = 0
On = 1

A single bit of information, then, is of very limited usefulness. But we can describe the state of the three-way lamp with 2 bits:

Off = 00
On = 01
Brighter = 10
Brightest = 11

Two bits allow us to communicate more information than we can with a single bit. In our lamp example, 2 bits allowed us to distinguish between four distinct states. The more bits we use in a unit, the more potential information we can express. A digital computer typically works with units of 8 bits (or multiples of 8, such as 16 and 32).

An 8-bit group is called a *byte.* A single byte allows all the letters, numbers, and other characters of a typewriter or computer keyboard to be encoded by a number, with room left over. Eight bits permits 256 different meanings to be given to a pattern of 1s and 0s. The number of different combinations or meanings for any given length of bits n is equal to 2 to the nth power, 2^n. For example, 16 bits yields 65,536 combinations, since $2^{16} = 65{,}536$. Each time a bit is added, the number of possible different values doubles.

A pulse train is often shown in its ideal form, as the string in Figure 2–4. The pulses go from one state to another instantaneously. Such diagrams show ideal pulses and the main essentials of the pulse train, so engineers and technicians can compare trains.

A pulse train represents the 1s and 0s of digital information. The train can depict high- and low-voltage levels or the presence and absence of a voltage. In electronics, a digital 1 can be a voltage pulse or a higher voltage level. A digital 0 can be the absence of the pulse or a lower voltage level. Thus we can speak of a 1 with terms such as *on* or *high* and of 0 with terms such as *off* or *low.*

A real pulse does not occur instantaneously as it is shown doing in Figure 2–4. An electronic circuit has a finite response time—that is, it takes time for a voltage or optical pulse to turn on and off or switch between high and low levels. The pulse must also stay on for a brief time. Still, in a computer system, turning a pulse on and off can require only millionths or billionths of a second, to allow thousands or millions of pulses per second.

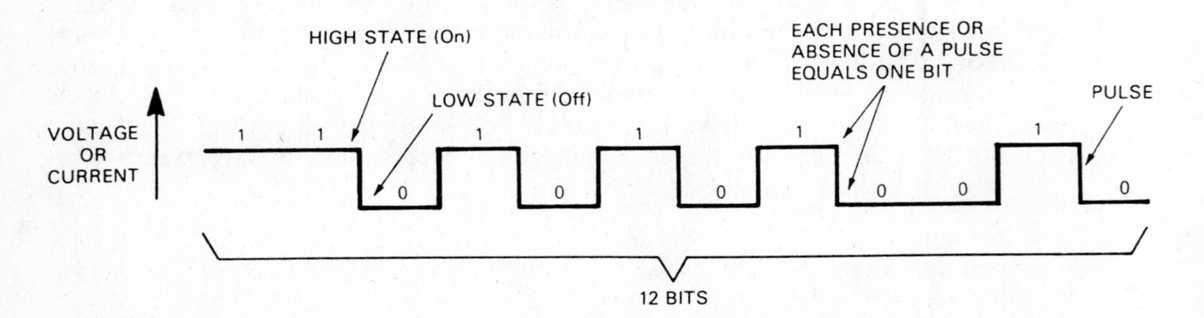

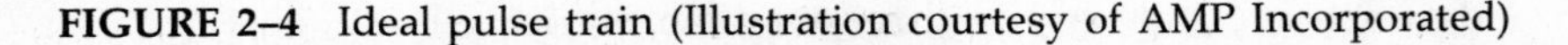
FIGURE 2–4 Ideal pulse train (Illustration courtesy of AMP Incorporated)

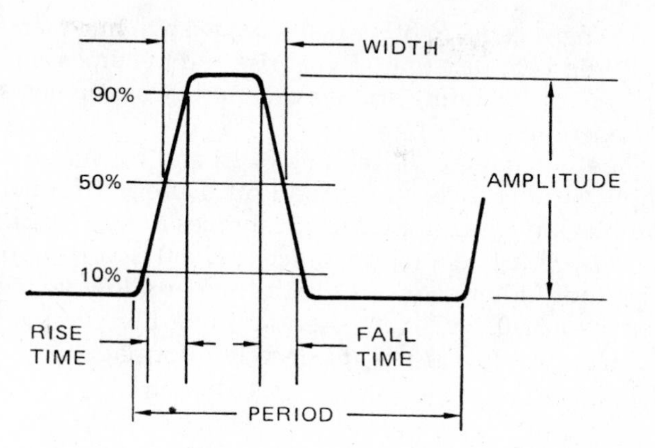

FIGURE 2–5 Parts of a pulse

Engineers dealing with digital systems must consider the shape of the pulse. Figure 2–5 defines the parts of the pulse.

- *Amplitude* is the height of the pulse. It defines the level of energy in the pulse. The energy can be voltage in a digital system or optical power in a fiber-optic system. Notice that different types of energy are used for different types of systems.
- *Rise time* is the time required for the pulse to turn on—to go from 10% to 90% of its maximum amplitude.
- *Fall time,* the opposite of rise time, is the time required to turn off, measured from 90% to 10% of amplitude. Rise time and fall time may or may not be equal.
- *Pulse width* is the width, expressed in time, of the pulse at 50% of the amplitude.
- *Bit period* is the time given for the pulse. Most digital systems are clocked or timed. Pulses must occur in the time allotted by the system for a bit period. For example, suppose a string of five 0s occurs. How do we know it is not four 0s or six 0s? We can detect no changes in the pulse train. We know because nothing has occurred in 5 bit periods of the clock.

A clock is an unchanging pulse train that provides timing by defining bit periods. Each bit period can be defined as one or more clock timing periods. The clock is loosely comparable to a musician's metronome: Both provide the basic timing beat against which operation (pulses or musical notes) is compared.

Rise time is especially important in electronics and fiber optics. Rise time limits the speed of the system. The speed at which pulses can be turned on and off will determine the fastest rate at which pulses can occur. The easiest way to increase the speed of a system is to decrease pulse rise and fall times, thereby turning pulses on and off faster. Thus more pulses can occur in a given time. Even when pulse amplitude and pulse width remain the same, faster rise times bring faster operating speeds. In other cases, we wish to maintain the pulse width as a certain proportion of pulse time. Faster rise and fall times allow shorter pulse widths, which increase operating speed even more. Conversely, slowing the rise time slows the operating speed.

As we will see in Chapter 10, the bits of 1s and 0s can be formed in other ways than simply turning a voltage on and off. Different formats for coding 1s and 0s offer various advantages and disadvantages in transmitting information. Here we have showed a 1 as a high voltage and a 0 as a low voltage occurring in a given bit period. Other codes use both high and low voltages within a bit period to encode a 1 or 0. Still, the essential thing here is that digital systems use 1s and 0s, high and low states, to encode information.

Why Digital?

Computers use digital signals because it is their nature: They are digital by design. Although a computer does not have to use digital techniques and binary numbers, it does so because they are the most convenient way to construct a computer that operates electronically. Transistors and integrated circuits made up of thousands of transistors operate as very fast on/off switches.

Telephone companies use digital techniques and turn your voice into a digital signal before transmitting it. What is the advantage? A signal transmitted any distance becomes distorted. Even if your voice is faithfully reproduced electronically for transmission as an analog signal, it still is distorted by the time it reaches the receiver. Unfortunately, the receiver cannot correct the distortion, because it has no way of knowing what the original signal looked like.

For digital pulses, the situation is reversed. A digital pulse has a defined shape. The receiver knows what the original pulse looked like. A receiver only needs to know information about the pulses: How many pulses? When were they sent? and so on. The receiver can be designed to rebuild distorted pulses into their original shape to faithfully reproduce the original signal.

INFORMATION-CARRYING CAPACITY

Any transmission path carrying signals has limits to the amount of information it can carry. The amount of information that the path can carry is its *information-carrying capacity*. Several ways exist to describe this capacity. In telephony, capacity is expressed in voice channels. A voice channel is the bandwidth or range of frequencies required to carry a single voice. Since the upper end of the human voice is about 4 kHz, a single voice channel must have a bandwidth of 4 kHz. In the early days of telephones, each wire carried only a single voice. Today, a single telephone line carries hundreds or thousands of voices simultaneously. It thus has a capacity of thousands of voice channels.

Bandwidth is proportional to the highest rate at which information can pass through it. If an optical fiber, for example, has a bandwidth of 400 MHz, it can carry frequencies up to this limit.

In digital systems, capacity is given in bits per second (bps) or *baud.* For a telephone system, a single digital voice channel requires 64,000 bps. A digital system, therefore, requires more bandwidth than a comparable analog system. An analog telephone system requires 4 kHz per voice; a digital system requires over 16 times more—64 kHz. Simple digital telephone systems, which carry 672 voices one way over a single line, have an operating speed of 44.7 megabits per second (Mbps).

The methods that allow such systems to be built are PCM and multiplexing.

PCM AND MULTIPLEXING

The technique used to turn your analog voice into a digital signal is *PCM.* Time-division multiplexing is the means that allows several voice channels to be transmitted on a single line.

Communication theory states that an analog signal, such as your voice, can be digitally encoded and decoded if it is sampled at twice the rate of its highest frequency. The high end of the speaking voice in telephony is 4000 Hz, which means that a PCM system must sample the voice 8000 times a second. *Sampling* means that the system looks at the amplitude of the vocal frequencies. Each sample is converted into an 8-bit number. Eight bits allow 256 different amplitudes to be encoded. Since each sample requires 8 bits and the sampling occurs 8000 times a second, the bandwidth required for a single voice channel is 64,000 bps (8000 samples $\times$ 8 bits/sample = 64,000 bits). At the receiver end, the same rules used to encode the sample are used to decode the sample and to reconstruct the analog voice signal. Figure 2–6 shows the idea of PCM. There are many ways to design and use PCM systems, but the basic idea is as we have described.

Sending 64,000 bps does not make efficient use of the capacity of a typical transmission path. To make full use of the capacity, telephone companies send several voice channels over the path. The appearance is that all channels are transmitted simultaneously, although they really are not. Part of one voice is first sent, and then part of the second, part of the third, and so on, in an interleaving pattern of channels. The device that permits this combination of different signals onto a single line is a *multiplexer.* The speed of the system easily allows all signals to be sent in turn. A *demultiplexer* at the receiver performs the opposite operation and separates the signals.

Time-division multiplexing (TDM) is the name given to the method of multiplexing that assigns parts of each voice channel to specific time slots. Other forms of multiplexing, such as *frequency-division multiplexing* (FDM) and *wavelength-division multiplexing* (WDM), exist. FDM assigns different information channels

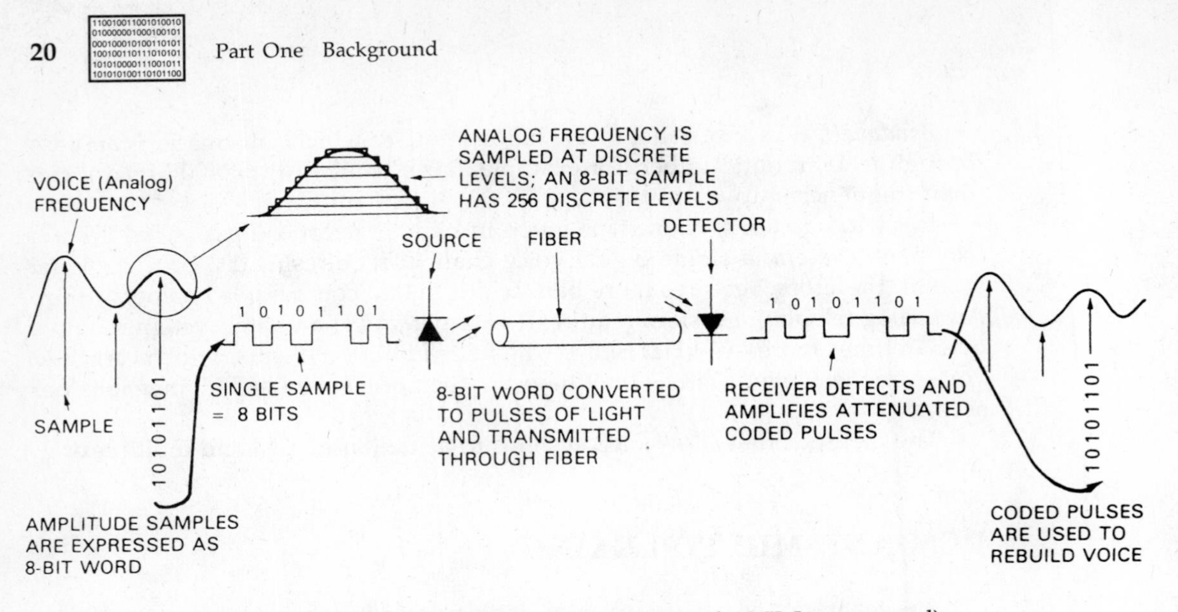

FIGURE 2–6 PCM (Illustration courtesy of AMP Incorporated)

to different carrier frequencies. Cable television uses FDM. WDM, which is unique to optical communications, is discussed in Chapter 12.

Fiber optics is important to such telephone systems because its information-carrying capacity exceeds wire-based systems.

THE DECIBEL

The *decibel* is an important unit that you will use continually in fiber optics as well as in electronics. It is used to express gain or loss in a system or component. A transistor, for example, can amplify a signal, making it stronger by increasing its voltage, current, or power. This increase is called *gain*. Similarly, loss is a decrease in voltage, current, or power. The basic equations for the decibel are

$$dB = 20 \log_{10} \left(\frac{V_1}{V_2} \right)$$

$$dB = 20 \log_{10} \left(\frac{I_1}{I_2} \right)$$

$$dB = 10 \log_{10} \left(\frac{P_1}{P_2} \right)$$

where V is voltage, I is current, and P is power. The decibel, then, is the ratio of two voltages, currents, or powers. Notice that voltage and current are 20 times the logarithmic ratio, and power is 10 times the ratio.

A basic use of the decibel is to compare the power entering a system, circuit, or component to the power leaving it. The decibel value shows how the device affects the circuit. A transistor, for example, usually increases the power. Other components may decrease the power. P_1 is the power out, and P_2 is the power in.

$$\text{dB} = 10 \log_{10} \left(\frac{P_{\text{out}}}{P_{\text{in}}} \right)$$

$$P_{\text{in}} \rightarrow \boxed{\text{circuit}} \rightarrow P_{\text{out}}$$

Another use of the decibel is to describe the effect of placing a component in a system. Sometimes, for example, it is necessary to split a cable and apply connectors so that the two cable halves can be connected and disconnected. Insertion of these connectors causes some loss, which is expressed in decibels. In other cases, devices called *attenuators* are purposely placed in a circuit to provide loss.

In fiber optics, we deal mostly with loss and optical power. (The electronic circuits in the transmitter and receiver may use voltage and current and provide gain.) The source emits optical power. As light travels through the fiber to the receiver, it loses power. This power loss is expressed in decibels. For example, if the source emits 1000 microwatts (μW) of power and the detector receives 20 μW, the loss through the system is about 17 dB:

$$\begin{aligned} \text{Loss} &= 10 \log_{10} \left(\frac{P_{\text{r}}}{P_{\text{tr}}} \right) \\ &= 10 \log_{10} \left(\frac{20}{1000} \right) \\ &= -16.989 \text{ dB} \end{aligned}$$

where P_{tr} is the power transmitted from the source and P_{r} is the power received by the receiver.

Table 2–1 shows the amount of power remaining for different decibel values. A 10-dB loss (–10 dB) represents a loss of 90% of the power; only 10% remains. An additional loss of 10 dB increases the loss by an order of magnitude. A useful figure to remember is 3 dB, which represents a loss of one half of the power.

Fiber-optic links easily tolerate losses of 30 dB, meaning that 99.9% of the power from the source is lost before it reaches the detector. If the source emits 1000 μW of power, only 1 μW reaches the detector. Of concern in fiber optics are the proper balance of the source power, the losses in the system, and the sensitivity of the detector to weak light signals.

Remember that a decibel expressing loss is a negative unit. In fiber optics, it is common practice to omit the negative sign and speak of a loss of 6 dB, say.

Loss (dB)	Power Remaining (%)	Loss (dB)	Power Remaining (%)
0.1	97.7	4	39.8
0.2	95.5	5	31.6
0.3	93.3	6	25.1
0.4	91.2	7	19.9
0.5	89.1	8	15.8
0.6	87.1	9	12.6
0.7	85.1	10	10.0
0.8	83.2	20	1.0
0.9	81.1	30	0.1
1	79.4	40	0.01
2	63.1	50	0.001
3	50.1	60	0.001

TABLE 2–1 Loss and the decibel

This loss is actually –6 dB. If you solved the equation, the result would be –6 dB. But in talking, and even in data sheets, the negative sign is omitted, with little confusion being caused. But if you have occasion to use a loss figure in an equation, do not forget to make the number negative! (The confusion arises because some equations have been adjusted to accept a loss figure as a positive number.)

Sometimes the ratio for calculating loss or gain uses a constant P_2. In fiber optics, this value is usually 1 milliwatt (mW).

$$\text{dBm} = 10 \log_{10} \left(\frac{P}{1 \text{ mW}} \right)$$

dBm means ''decibels referenced to a milliwatt.'' In this case, the negative sign is almost always used. A value of –10 dBm means that P is 10 dB less than 1 mW, or 100 μW. Similarly, –3 dBm is 500 μW. Communication engineers and technicians use dBm units extensively.

Figure 2–7 relates various dBm values to milliwatt and microwatt levels. Notice that a small range of the dBm values allows a great range of power values to be expressed.

Another unit sometimes used is dBμ—decibel referenced to 1 μW. This is the same as dBm, except the reference value is 1 μW rather than 1 mW.

$$\text{dB}\mu = 10 \log_{10} \left(\frac{P}{1\ \mu\text{W}} \right)$$

10 mW	=	+10 dBm
5 mW	=	+ 7 dBm
1 mW	=	0 dBm
500 μW	=	− 3 dBm
100 μW	=	−10 dBm
50 μW	=	−13 dBm
10 μW	=	−20 dBm
5 μW	=	−23 dBm
1 μW	=	−30 dBm
100 nW	=	−40 dBm
10 nW	=	−50 dBm
1 nW	=	−60 dBm
100 pW	=	−70 dBm
10 pW	=	−80 dBm
1 pW	=	−90 dBm

FIGURE 2–7 Power-to-dBm conversion

SUMMARY

- Communication consists of encoding, transmitting, and decoding information.
- Digital systems work with bits representing 1 and 0.
- Information-carrying capacity expresses the amount of information that can be transmitted over a communication path.
- PCM turns an analog signal into a digital signal.
- Multiplexing allows many signals to be sent over a single communications path.
- The decibel is the primary unit for indicating gain or loss in a system.

? REVIEW QUESTIONS

1. Name the three activities involved in communication. Give an example of each.
2. Define modulation.
3. Sketch an analog signal. Sketch a digital signal.
4. Is the information 1101101 analog or digital?
5. Sketch a pulse. Label its amplitude, rise time, fall time, and pulse width.
6. What is the capacity of an analog voice channel? Of a digital voice channel?
7. For a PCM system as described in this chapter, how many bits per second are required to transmit five voice channels? 50 voice channels?
8. Name the technique that allows five voice channels to be transmitted simultaneously over a single optical fiber.
9. Which requires greater bandwidth, an analog voice channel or a digital voice channel?
10. How many milliwatts are in a signal having a power of 0 dBm?

CHAPTER 3 Fiber Optics as a Communications Medium: Its Advantages

In its simplest terms, fiber optics is a communications medium linking two electronic circuits. The fiber-optic link may be between a computer and its peripherals, between two telephone switching offices, or between a machine and its controller in an automated factory. Obvious questions concerning fiber optics are these: Why go to all the trouble of converting the signal to light and back? Why not just use wire? The answers lie in the following advantages of fiber optics.

- Wide bandwidth
- Low loss
- Electromagnetic immunity
- Light weight
- Small size
- Safety
- Security

The importance of each advantage depends on the application. In some cases, wide bandwidth and low loss are overriding factors. In others, safety and security are the factors that lead to the use of fiber optics. The following pages discuss each advantage in detail.

WIDE BANDWIDTH

The last chapters showed the increasing concern with sending more and more information electronically. Potential information-carrying capacity increases with the bandwidth of the transmission medium and with the frequency of the carrier. From the earliest days of radio, useful transmission frequencies have pushed upward five orders of magnitude, from about 100 kHz to about 10 GHz. The frequencies of light are several orders of magnitude above the highest radio waves. The invention of the laser, which uses light as a carrier, in a single step increased the potential range four more orders of magnitude—to 100,000 GHz (or 100 terahertz [THz]). Optical fibers have a potential useful range to about 1 THz, although this range is far from being exploited today. Still, the practical bandwidth of optical fibers exceeds that of copper cables. Furthermore, the information-carrying possibilities of fiber optics have only begun to be exploited, whereas the same potentials of copper cable are pushing their limits.

As mentioned earlier, telephone companies increasingly use digital transmission. The higher bandwidth of optics allows a higher bit rate and, consequently, more voice channels per cable. For compatibility to exist among all telephone carriers, the rates at which different transmission lines carry information are generally fixed by a system known as the North American Digital Telephone Hierarchy.

Table 3–1 shows the hierarchy for coaxial cables and optical fibers. The coaxial system has been long established. The fiber-optic capacities shown are for Sonet, or synchronous optical network. Chapter 15 describes Sonet in greater detail.

One consequence of fiber's high bandwidth is that it permits transmission of channels requiring much greater bandwidth than a voice channel. Television and teleconferencing, for example, require a channel capacity 14 to 100 times that of a digitally encoded voice. The bandwidth of a fiber allows these signals to be multiplexed through the fiber, permitting voice, data, and video to be transmitted simultaneously. The demand for these services means that fibers will move from being only long-distance carriers to being carriers right to the home and business.

To give perspective to the incredible capacity that fibers are moving toward, a 10-Gbps signal has the ability to transmit any of the following *per second:*

- 1000 books
- 130,000 voice channels

Medium	Designation	Bit Rate (Mbps)	Voice Channels	Repeater Spacing (km)
Coaxial Cable	DS-1	1.544	24	1–2
	DS-1C	3.152	48	
	DS-2	6.312	96	
	DS-3	44.736	672	
Fiber (Sonet)	OC-1	51.84	672	25 (Laser)
	OC-3	155.52	2016	2 (LED)
	OC-9	466.56	6048	
	OC-12	622.08	8064	
	OC-18	933.12	12,096	
	OC-24	1244.16	16,128	
	OC-36	1866.24	24,192	
	OC-48	2488.32	32,256	
	OC-96	4976.64	64,512	
	OC-192	9953.28	129,024	

TABLE 3–1 Digital telephone transmission rates

- 16 high-definition television (HDTV) channels or 100 HDTV channels using compression techniques. (An HDTV channel requires a much higher bandwidth than today's standard television.)

Similar advances will not occur for coaxial systems: Optical systems will become the favored solution for long-distance, high-data-rate applications.

Another consequence of the high bandwidth of fiber optics is that it permits transmission of channels requiring much greater bandwidth than a voice. Television and teleconferencing require a channel capacity 14 to 100 times that of a digitally encoded voice. The bandwidth of the fiber allows these signals also to be multiplexed through the fiber. Fibers allow telephone companies to simultaneously transmit voice, data, and video, which is important to the information age.

LOW LOSS

Bandwidth is an effective indication of the rate at which information can be sent. Loss indicates how far the information can be sent. As a signal travels along a transmission path, be it copper or fiber, the signal loses strength. This loss of strength is called *attenuation.* In a copper cable, attenuation increases with modulation frequency: The higher the frequency of the information signal, the greater the loss. In an optical fiber, attenuation is flat; loss is the same at any signaling frequency up until a very high frequency. Thus, the problem of loss becomes greater in a copper cable as information-carrying capacity increases.

Figure 3–1, by showing the loss characteristics for fibers, twisted pairs, and coaxial cable, such as are used in telephone systems, demonstrates the usable ranges for these transmission media. Loss in coaxial cable and twisted-pair wires increases with frequency, whereas loss in the optical cable remains flat over a very wide range of frequencies. The loss at very high frequencies in the optical fiber does not result from additional attenuation of the light itself by the fiber. This attenuation remains the same. The loss is caused by loss of information, not by optical power. Information is contained in the variation of the optical power. At very high frequencies, distortion causes a reduction or loss of this information.

The point is that the effects of loss that must be accounted for in a system depend on the signal frequency. What is suitable for a system working at one speed may not work at another frequency. This need to consider different signaling speeds complicates designs. High-frequency design, for example, is more difficult than low-frequency design. One cannot simply increase operating speeds without taking into account the effects of speed on performance. In a fiber-optic system, the loss is the same at all operating speeds within the fiber's specified bandwidth.

Severe attenuation requires repeaters at intermediate points in the transmission path. For copper cables, repeater spacings, in general, decrease as operating speeds increase. For fibers, the opposite is true: Repeater spacings increase along with operating speeds, because at high data rates very efficient, low-loss fibers are used.

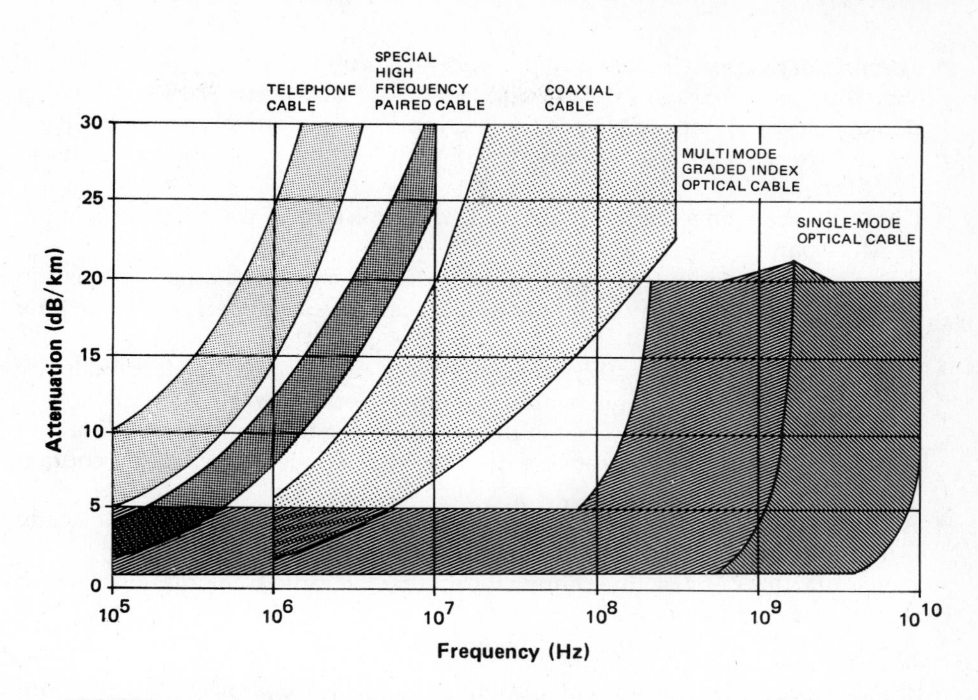

FIGURE 3–1 Attenuation versus frequency (Courtesy of Siecor Corporation)

The first transatlantic fiber-optic telephone link, installed by AT&T in 1988, carried 37,800 simultaneous voice conversations each direction on a pair of fibers. Repeater spacings are 35 km. In contrast, the best transatlantic coaxial system handles 4200 conversations and has a repeater spacing of only 9.4 km. The possibility exists that fiber-optic systems will carry 200 Mbps over 80 to 100 km without a repeater.

The combination of high bandwidth and low loss has made the telephone industry probably the heaviest user of fiber optics. It allows them not only to send more information over a transmission path but also to extend repeater spacings. Repeaters, after all, are electronics systems that are costly to build, install, and maintain. Fewer repeaters mean less costly systems.

ELECTROMAGNETIC IMMUNITY

Unlike copper cables, optical fibers do not radiate or pick up electromagnetic radiation. Any copper conductor acts like an antenna, either transmitting or receiving energy. One piece of electronic equipment can emit electromagnetic interference (EMI) that disrupts other equipment. Among reported problems resulting from EMI are the following:

- The military reported such a high concentration of electronic equipment in Vietnam that increases in the population of electronic devices beyond a crucial density made it impossible for the equipment to operate.
- An electronic cash register interfered with aeronautical transmissions at 113 MHz.
- Coin-operated video games interfered with police radio transmissions in the 42-MHz band.
- Some personal computers tested by the Federal Communications Commission (FCC) in 1979 emitted enough radiation to disrupt television reception several hundred feet away.
- Electrostatic discharges (ESD), discharged into computer terminals by operators, have garbled the computer memory, erased work in progress, and even destroyed circuits. (ESD is the shock you get when you walk across a carpet on a dry day and touch a doorknob. Such discharges easily contain 15 to 25 kV.)
- An explosion caused by static electricity killed three workers at Cape Kennedy in the 1960s.
- A manufacturer of gasoline pumps found that CB radio transmissions zeroed remote pump readouts.
- Airport radar erased taxpayers' records in a computer memory bank.

EMI is a form of environmental pollution with consequences ranging from the merely irksome to the deadly serious. As the density of electronic equipment grows (and it will as part of the information age), the potential for EMI problems also increases. To combat such problems, the FCC in 1979, for example, issued stringent regulations limiting EMI from computing devices. Regulatory agencies in Europe have issued similar regulations.

Here is a simple way to demonstrate the effects of EMI if you have access to a personal computer or computer terminal. Place an AM radio near the computer while it is running. Tune through the dial. At some point, the radio should pick up the computer, so that you will be able to listen to the computer operate. Try running different programs to hear differences. What you hear is EMI.

Cables interconnecting equipment can be one of the main sources of EMI. They can also be one of the main receiving antennas carrying EMI into equipment. Cables act just like the radio antenna in the foregoing EMI demonstration.

Since fibers do not radiate or receive electromagnetic energy, they make an ideal transmission medium when EMI is a concern. Some factories use fiber optics because of its immunity. These applications do not require the high bandwidth or low loss of fibers. Equipment such as motors that switch on and off can be a source of EMI that disrupts signal lines controlling the equipment. Using fibers rather than copper cables eliminates the problems.

High-voltage lines can present another problem, since they also emit energy. Copper signal cables cannot be run next to such lines without special precautions, because energy from the high-voltage line couples onto the signal line. Fiber-optic lines can be run beside high-voltage lines with no detrimental effect, since no energy couples onto it from the high-voltage line.

One consequence of the fiber's electromagnetic immunity is that signals do not become distorted by EMI. Digital transmission requires that signals be transmitted without error. EMI can be a cause of error in electrical conductor transmission systems. A burst of EMI may appear as a pulse, where no pulse occurred in the original pulse stream. Fibers offer very high standards in error-free transmission.

LIGHT WEIGHT

A glass fiber weighs considerably less than a copper conductor. A fiber-optic cable with the same information-carrying capacity as a copper cable weighs less than the copper cable because the copper requires more lines than the fiber. For example, a typical single-conductor fiber-optic cable weighs 9 lb/1000 ft. A comparable coaxial cable weighs nine times as much—about 80 lb/1000 ft. Weight savings are important in such applications as aircraft and automobiles.

SMALL SIZE

A fiber-optic cable is smaller than its copper counterparts. In addition, a single fiber can often replace several copper conductors. Figure 3–2 shows a comparison

FIGURE 3–2 Size comparison: coaxial cable and fiber-optic cable (Courtesy of AT&T Bell Laboratories)

of coaxial cable and a fiber-optic cable used in digital telephony. The copper cable is 4.5 in. in diameter and can carry as many as 40,300 two-way conversations over short distances. The fiber-optic cable, which contains 144 fibers in its 0.5-in.-diameter structure, has the capacity to carry 24,192 conversations on each fiber pair or nearly 1.75 million calls on all the fibers. The fiber-optic cable greatly exceeds the capacity of the coaxial cable even though it is almost 10 times smaller.

The small size of fiber-optic cables makes them attractive for applications where space is at a premium:

- Aircraft and submarines, where the use of every square inch is critical. Not only do the fiber-optic cables use less of the valuable space, they can be placed in areas where copper cables cannot be. In short, they make efficient use of the space.
- Underground telephone conduits, especially in cities. Here, not only are conduits often filled to capacity, but building new conduits to allow expanded services is very costly. Additional space may be unavailable. Fiber-optic cables replace the copper cables, often offering greater capacity in less space. A large copper cable that fills a conduit can be replaced by a smaller fiber cable, with room left for new cables in the future.
- Computer rooms, where the cables between equipment run under raised floors. These cables are often very rigid and difficult to install. Adding new cables is also difficult. Again, the small size and resulting flexibility of fiber eliminate such problems. Indeed, in some cases, so few fibers are needed that the need for a raised floor can also be eliminated.

SAFETY

A fiber is a dielectric—it does not carry electricity. It presents no spark or fire hazard, so it cannot cause explosions or fires as a faulty copper cable can. Furthermore, it does not attract lightning. The fiber-optic cable can be run through hazardous areas, where electrical codes or common sense precludes the use of wires. It is possible, for example, to run a fiber directly through a fuel tank.

SECURITY

One way to eavesdrop is to tap a wire. Another way is to pick up energy radiated from a wire or equipment (a form of EMI). Years ago, the United States discovered a foreign power doing just that to our embassies. A sensitive antenna in a nearby building was secretly picking up energy radiated from electronic equipment in the embassy. The antenna was receiving EMI much like the radio example discussed earlier. This energy, though, included top-secret and classified data. Businesses also spend millions of dollars each year protecting their secrets, such as encrypting data before it is transmitted.

Fiber optics is a highly secure transmission medium. It does not radiate energy that can be received by a nearby antenna, and it is extremely difficult to tap a fiber. Both government and business consider fiber optics a secure medium.

CONCLUSION

High bandwidth, low loss, and electromagnetic immunity are probably the three most outstanding features of fiber optics. These features dovetail nicely. They allow high-speed data transfer over long distances with little error. A fiber-optic link is capable of transmitting the entire text of a 30-volume encyclopedia over 100 miles in 1 s. The level of error is only one or two incorrect letters during the transmission.

Realize, however, that not all fibers have low loss and high bandwidth. Where loss and high speeds are not critical, as in automobiles, for example, less-expensive fibers work well. In an automobile, the central concern is protecting against noise from sources such as the ignition system. The other advantageous features of fiber optics make a well-suited transmission medium in many applications.

SUMMARY

- Fiber optics offers many advantages over copper cables.
- Optical fibers offer greater bandwidth.
  Optical fibers offer lower losses.
- High bandwidth and low losses mean greater repeater spacing in long-distance systems.
- Fibers offer excellent protection against EMI.
- Fibers are a secure transmission medium.
- Fibers are smaller and lighter than comparable copper cables.
- Since fibers do not carry electricity and cannot spark or cause fires, they are safe even in hostile environments.
- Fibers are used for different reasons, such as low loss, high bandwidth, security, and EMI immunity.

? REVIEW QUESTIONS

1. List six advantages of fiber optics. Give an example of each.
2. Give three examples of rates and capacities in the North American Digital Telephone Hierarchy.

3. Which of the following is the most important reason for large bandwidth in optical fibers:
   A. High-speed, high-capacity transmission
   B. Secure transmission
   C. Fewer repeaters
   D. Immunity from electromagnetic interference
4. Does loss increase, decrease, or stay the same as the signal frequency increases in a copper cable? In an optical cable?
5. What is the name of the phenomenon that causes ghosts on your television, crackles on your radio, and other malfunctions in electronic equipment? Why does an optical fiber not contribute to this phenomenon?
6. For a 200-kHz signal traveling 50 m through a factory, which of the following factors are of great importance? Of little importance? Explain your reasons.
   A. Bandwidth
   B. Low loss
   C. EMI immunity
   D. Safety
   E. Error-free transmission
7. The original interest in optical fibers as a communication medium arose because
   A. They used less space than a coaxial cable.
   B. They were less expensive than copper.
   C. They could carry the high frequencies of laser light with little loss.
   D. Their electromagnetic immunity, security, and compatibility with digital techniques made them of use to the military.
8. (True/False) Information-carrying capacity depends on the physical size of the medium carrying the capacity.

Light

Light is electromagnetic energy, as are radio waves, radar, television and radio signals, x-rays, and electronic digital pulses. *Electromagnetic energy* is radiant energy that travels through free space at about 300,000 km/s or 186,000 miles/s. An electromagnetic wave consists of oscillating electric and magnetic fields at right angles to each other and to the direction of propagation. Thus, an electromagnetic wave is usually depicted as a sine wave, as shown in Figure 4–1.

The main distinction between different waves lies in their frequency or wavelength. Frequency, of course, defines the number of sine-wave cycles per second and is expressed in hertz. *Wavelength* is the distance between the same points on two consecutive waves (or it is the distance a wave travels in a single cycle). Wavelength and frequency are related. Wavelength (lambda) equals the velocity of the wave (v) divided by its frequency (f):

$$\lambda = \frac{v}{f}$$

In free space or air, the velocity of an electromagnetic wave is the speed of light.

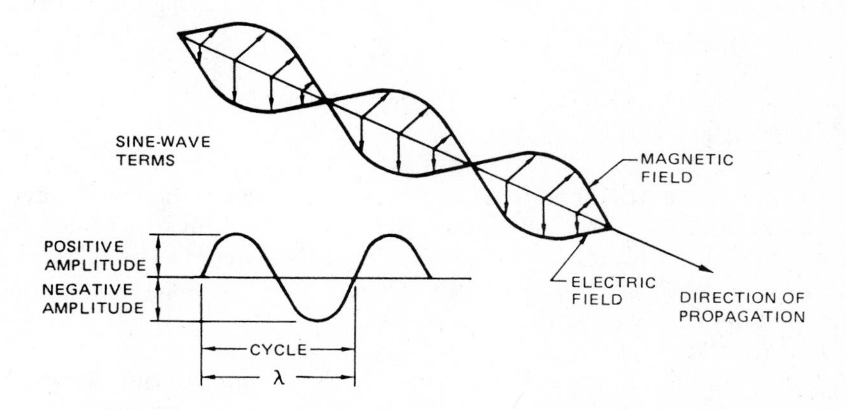

FIGURE 4–1 An electromagnetic wave

The equation clearly shows that the higher the frequency, the shorter the wavelength. For example, the 60-Hz power delivered to your house has a wavelength of 3100 miles. A 55.25-MHz signal, which carries the picture for channel 2 on television, has a wavelength of 17.8 ft. Deep red light has a frequency of 430 THz (430×10^{12} Hz) and a wavelength of only 700 nm (nanometers or billionths of a meter).

In electronics, we customarily talk in terms of frequency. In fiber optics, however, light is described by wavelengths. Remember, though, that frequency and wavelength are inversely related.

THE ELECTROMAGNETIC SPECTRUM

Electromagnetic energy exists in a continuous spectrum of frequencies from subsonic energy through radio waves, microwaves, γ (gamma) rays, and beyond. (Electromagnetic energy exists in the subsonic range of the spectrum; however, sound itself is not electromagnetic energy but a vibration of air molecules.) Figure 4–2 shows the spectrum. Notice that the radio frequencies most commonly used for communication are well below those of light.

Thus, light is electromagnetic energy with a *higher* frequency and *shorter* wavelength than radio waves. The figure also shows that the visible light we can see is only a small part of the light spectrum. Visible light has wavelengths from 380 nm for deep violet to 750 nm for deep red. Infrared light has longer wavelengths (lower frequencies) than visible light, whereas ultraviolet light has shorter ones. Most fiber-optic systems use infrared light between 800 and 1500 nm because glass fibers carry infrared light more efficiently than visible light.

The high frequencies of light have made it of such interest to communications engineers. As we saw in Chapter 2, a higher-frequency carrier means greater information-carrying capacity. Fiber optics is a method of using this information-carrying potential of light.

WAVES AND PARTICLES

So far, we have described light as an oscillating electromagnetic wave. It is spread out in space, without a definite, discrete location. Physicists once divided all matter into either waves or particles. We usually think of light as a wave and an electron as a particle. Modern physicists, however, have shown that this distinction does not exist. Both light and electrons exhibit wavelike and particle-like characteristics.

A particle of light is called a *photon,* which is a *quantum* or bundle of energy. A quantum exists in fixed discrete units of energy—you cannot have half a quantum or 5.33 quanta. The amount of energy possessed by a photon depends

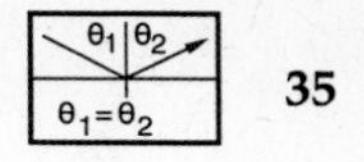

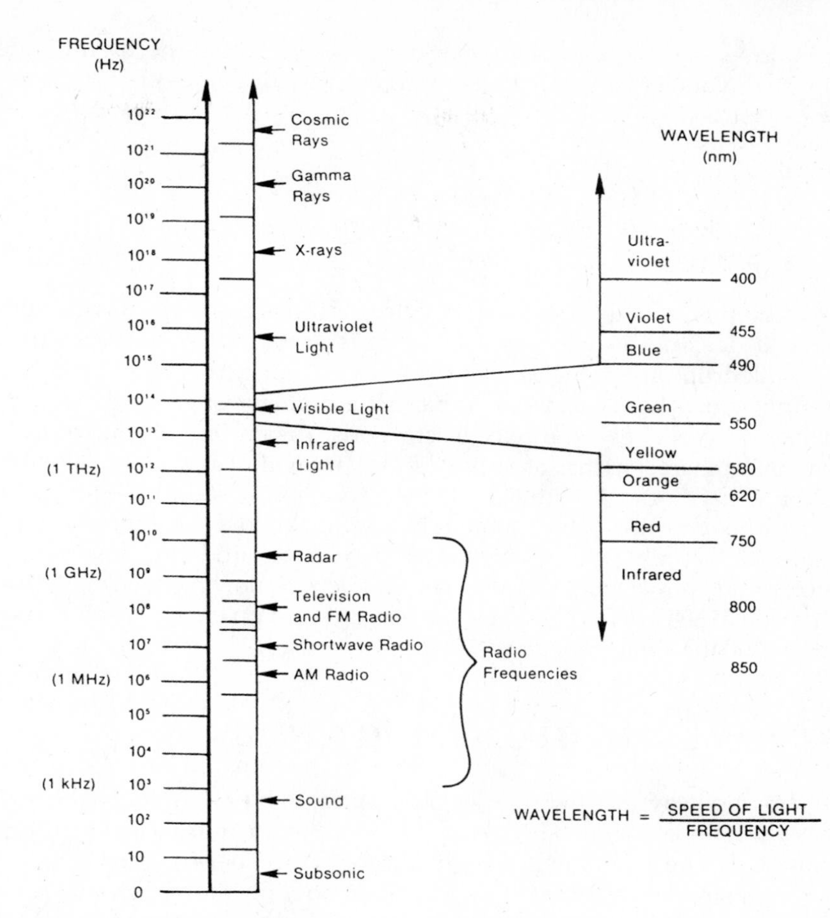

FIGURE 4–2 The electromagnetic spectrum (Illustration courtesy of AMP Incorporated)

on its frequency. The amount of energy increases as frequency increases: Higher frequency means more energy. Wavelengths of violet light have more energy than those of red light because they have higher frequencies. The energy *E*, in watts, contained in a photon is

$$E = hf$$

where f is its frequency and h is Planck's constant, which is 6.63×10^{-34} J-s (joule-seconds). The equation shows clearly that differences in photon energy are strictly functions of frequency (or wavelength). Photon energy is proportional to frequency. A photon is a quantum of light energy hf.

Here are some approximate energy levels for different high-frequency wavelengths. Notice that the higher the frequency, the more energy contained in the quantum.

Infrared light (10^{13} Hz):	6.63×10^{-20} J-s
Visible light (10^{14} Hz):	6.63×10^{-19} J-s
Ultraviolet light (10^{15} Hz):	6.63×10^{-18} J-s
X-rays (10^{18} Hz):	6.63×10^{-15} J-s

The photon is actually a strange particle, for it has zero rest mass. If it is not in motion, it does not exist! In this sense, it is not a particle as, say, marbles, stones, or ink drops are particles. It is a bundle of energy that acts like a particle.

Treating light as both a wave and a particle aids in investigation of fiber optics. We switch easily between the two descriptions, depending on our needs. For example, many characteristics of optical fibers vary with wavelength, so the wave description is used. On the other hand, the emission of light by a source or its absorption by a detector is best treated by particle theory. The description of a detector speaks of light photons striking the detector and being absorbed. This absorption provides the energy required to set electrons flowing as current. A light-emitting diode (LED) operates because its electrons give up energy as photons. The exact energy of the photon determines the wavelength of the emitted light.

LIGHT RAYS AND GEOMETRIC OPTICS

The simplest way to view light in fiber optics is by ray theory. The light is treated as a simple ray, shown by a line. An arrow on the line shows the direction of propagation. The movement of light through the fiber-optic system can be analyzed with simple geometry. This approach not only simplifies analysis, but it also makes the operation of an optical fiber simple to understand.

REFLECTION AND REFRACTION

What is commonly called the *speed of light* is actually the velocity of electromagnetic energy in a vacuum such as space. Light travels at slower velocities in other materials such as glass. Light traveling from one material to another changes speed, which, because of wave motion, results in light changing its direction of travel. This deflection of light is called *refraction.* In addition, different wavelengths of light travel at different speeds in the *same* material. The variation of velocity with wavelength plays an important role in fiber optics.

People who fish are quite familiar with refraction, since it distorts the apparent position of a fish lying stationary underwater (Figure 4–3). If one stands on a pier and looks directly down on the fish, the light is not refracted so that the fish is

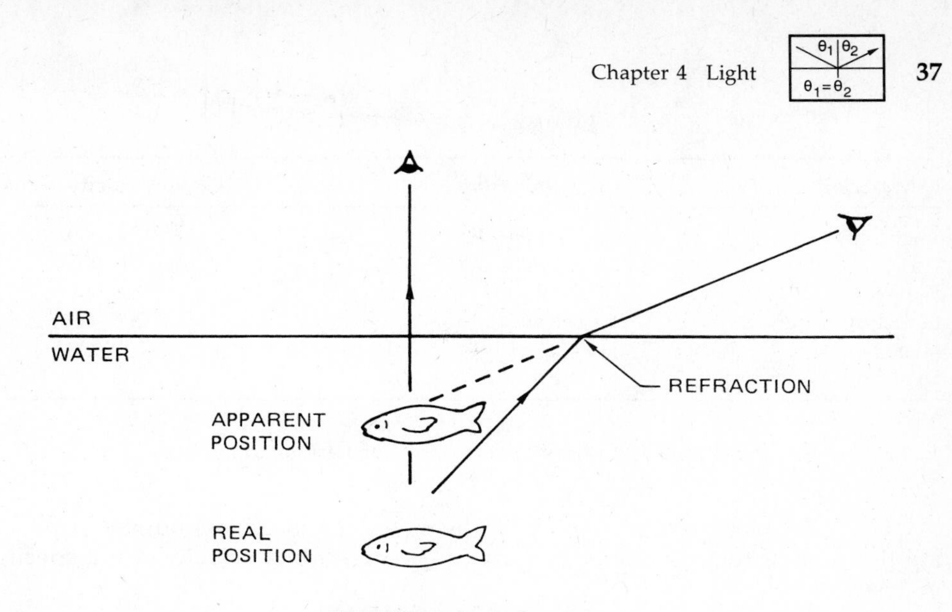

FIGURE 4–3 Refraction

where it appears to be. If the fish is viewed by looking into the water at an angle, refraction of light occurs. What appears to be a straight line from the fish to the eye is actually a line with a bend where the light passes from water into air and is refracted. As a result, the fish is actually deeper in the water than it appears to be.

The prism in Figure 4–4 also demonstrates refraction. White light entering the prism contains all colors. The prism refracts the light, and it changes speed as it enters the prism. Because each color or frequency changes speed differently, each is refracted differently. Red light deviates the least and travels the fastest. Violet light deviates the most and travels the slowest. The light emerges from the prism divided into the colors of the rainbow. Notice that refraction occurs at the entrance and at the exit of the prism.

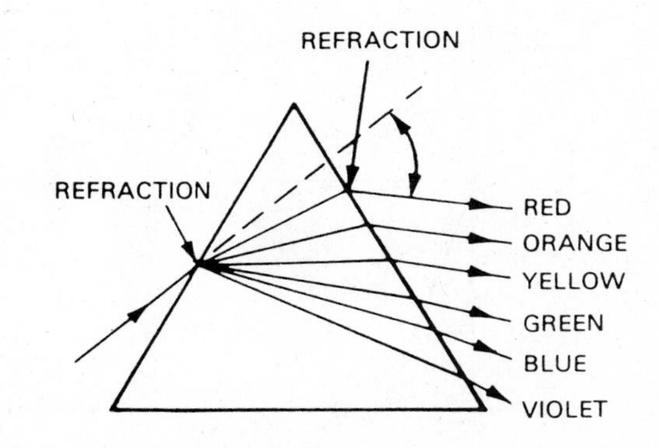

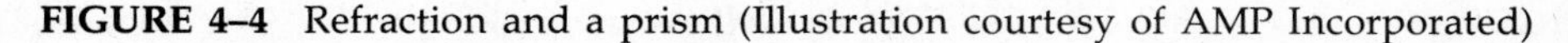

FIGURE 4–4 Refraction and a prism (Illustration courtesy of AMP Incorporated)

Material	Index (n)	Light Velocity (km/s)
Vacuum	1.0	300,000
Air	1.0003 (1)	300,000
Water	1.33	225,000
Fused quartz	1.46	205,000
Glass	1.5	200,000
Diamond	2.5	120,000

TABLE 4–1 Indices of refraction for various materials

The index of refraction, symbolized by n, is a dimensionless number expressing the ratio of the velocity of light c in free space to its velocity v in a specific material:

$$n = \frac{c}{v}$$

Table 4–1 lists some representative indices of refractions for selected materials as well as the approximate speed of light through the materials.

Of particular importance to fiber optics is that the glass's index of refraction can be changed by controlling its composition.

The amount that a ray of light is refracted depends on the refractive indices of the two materials. But before looking at the mechanics of refraction, we must first define some terms vital to our discussion. Figure 4–5 illustrates several important terms to understand light and its refraction.

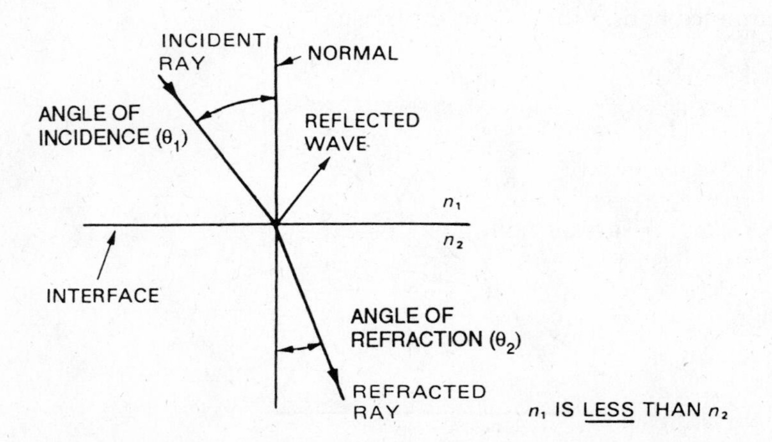

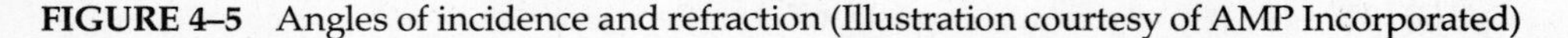

FIGURE 4–5 Angles of incidence and refraction (Illustration courtesy of AMP Incorporated)

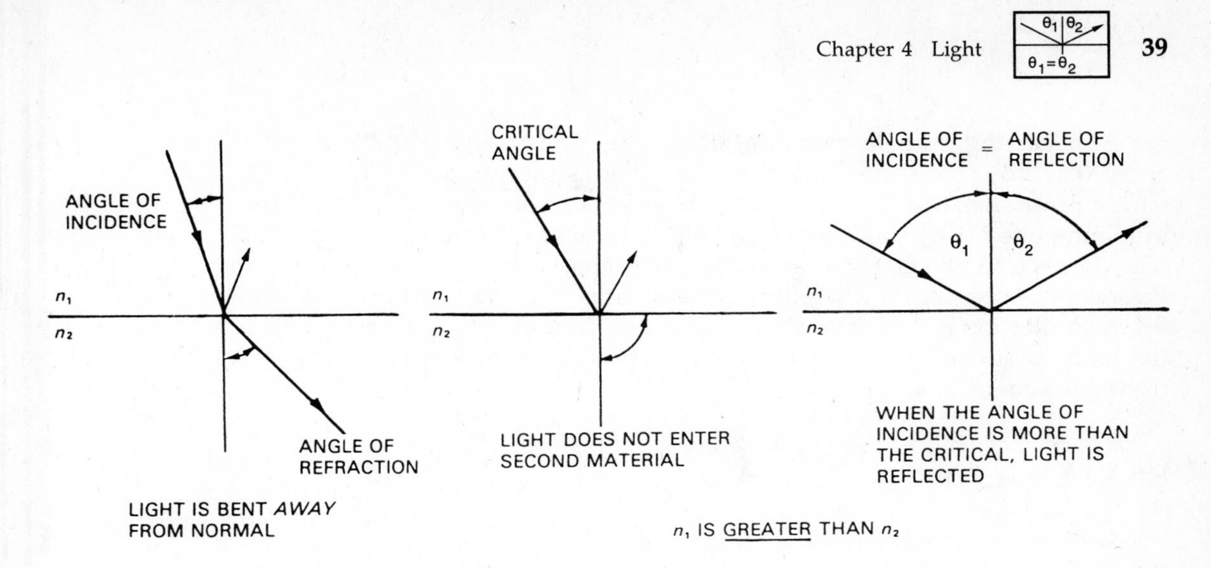

FIGURE 4–6 Reflection (Illustration courtesy of AMP Incorporated)

- The *normal* is an imaginary line perpendicular to the interface of the two materials.
- The *angle of incidence* is the angle between the incident ray and the normal.
- The *angle of refraction* is the angle between the refracted ray and the normal.

Light passing from a lower refractive index to a higher one is bent *toward* the normal. But light going from a higher index to a lower one refracts *away* from the normal, as shown in Figure 4–6. As the angle of incidence increases, the angle of refraction approaches 90° to the normal. The angle of incidence that yields an angle of refraction of 90° is the *critical* angle. If the angle of incidence increases past the critical angle, the light is totally reflected back into the first material so that it does not enter the second material. The angles of incidence and reflection are equal.

FRESNEL REFLECTIONS

Even when light passes from one index to another, a small portion is always reflected back into the first material. These reflections are *Fresnel reflections.* A greater difference in the indices of the materials results in a greater portion of the light reflecting. Fresnel reflection (ρ) at the boundary between air and another equals

$$\rho = \left(\frac{n - 1}{n + 1}\right)^2$$

In decibels, this loss of transmitted light is

$$\text{dB} = 10 \log_{10} (1 - \rho)$$

For light passing from air to glass (with $n = 1.5$ for glass), Fresnel reflection is about 0.17 dB. This figure will vary somewhat as the composition of the glass varies. Since such losses occur when light enters or exits an optical fiber, the loss in joining one fiber to another is 0.34 dB. A Fresnel reflection occurs when the light passes from the first fiber into the air gap separating the two fibers. A second Fresnel reflection occurs when the light passes from the air into the second fiber. Fresnel reflection is the same regardless of the order of materials through which the light passes; in other words, Fresnel reflection is the same as light passes from glass to air or from air to glass.

SNELL'S LAW

Snell's law states the relationship between the incident and refracted rays:

$$n_1 \sin \theta_1 = n_2 \sin \theta_2$$

where θ_1 and θ_2 are defined in Figures 4–5 and 4–6.

The law shows that the angles depend on the refracted indices of the two materials. Knowing any three of the values, of course, allows us to calculate the fourth through simple rearrangement of the equation.

The critical angle of incidence θ_c, where $\theta_2 = 90°$, is

$$\theta_c = \arcsin \left(\frac{n_2}{n_1} \right)$$

At angles greater than θ_c, the light is reflected. Because reflected light means that n_1 and n_2 are equal (since they are in the same material), θ_1 and θ_2 are also equal. The angles of incidence and reflection are equal. These simple principles of refraction and reflection form the basis of light propagation through an optical fiber.

A PRACTICAL EXAMPLE

Figure 4–7 shows an example of reflection that has practical application in fiber optics. Assume we have two layers of glass, as shown in Figure 4–7A. The first layer, n_1, has a refractive index of 1.48; the second, 1.46. These values are typical for optical fibers. Using Snell's law, we can calculate the critical angle:

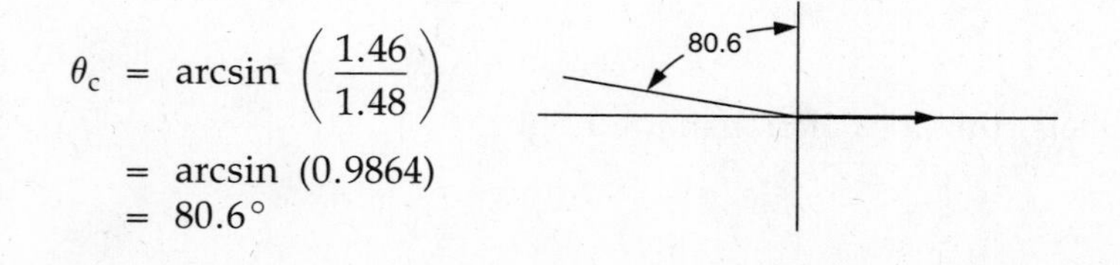

$$\theta_c = \arcsin \left(\frac{1.46}{1.48} \right) = \arcsin (0.9864) = 80.6°$$

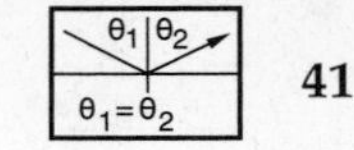

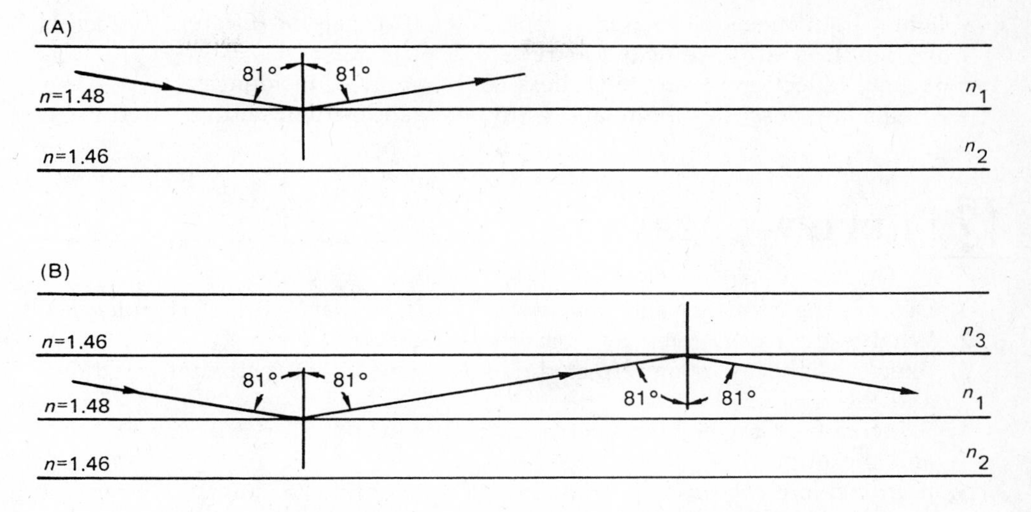

FIGURE 4–7 A practical example of reflection

Light striking the boundary between n_1 and n_2 at an angle greater than 80.6° from the normal reflects back into the first material. Again, the angle of incidence equals the angle of reflection.

Figure 4–7B carries the example a step further. Assume that a third layer of glass—labeled n_3—with the same refractive index as n_2 is placed on top of material n_1. Material n_1 is sandwiched between the n_2 and n_3 materials. We again have the same boundary condition as before. The reflected ray, however, now becomes the incident ray at the new boundary. The critical angle remains 80.6°. Thus all conditions are the same as in the first reflection. As a result, the light is again reflected back into the first material. The light reflected from n_3 becomes the incident ray again for n_2. The situation is the same. We have trapped the light between the two layers n_2 and n_3. As long as the angle of incidence is greater than 80.6°, the light is reflected back into the first material. Snell's law shows that, for this ideal example, this will always hold true. The light will continue traveling through the first material by total internal reflection.

The same principle explains the operation of an optical fiber. The main difference is that the fiber uses a circular configuration, such that n_2 surrounds n_1. The next chapter looks at light propagation in a fiber in detail.

SUMMARY

- Light is electromagnetic energy with a higher frequency and shorter wavelength than radio waves.
- Light has both wavelike and particlelike characteristics.

- When light meets a boundary separating materials of different refractive indices, it is either refracted or reflected.
- Fresnel reflections occur regardless of the angle of incidence.
- Snell's law describes the relationship between incident and reflected light.

 REVIEW QUESTIONS

1. Describe the frequency and wavelength of light in relation to radio frequencies.
2. What is the name given to a particle of light?
3. Sketch a light ray being refracted. Label the incident ray, the refracted ray, and the normal.
4. Sketch a light ray being reflected. Label the incident ray, the reflected ray, and the normal.
5. If an incident ray passing from air into water has a 75° angle with respect to the normal, what is the angle of refracted ray?
6. What is the wavelength of a 300-MHz electromagnetic wave in free space? A 250-kHz wave? A 2-GHz wave?
7. Define Fresnel reflection.
8. Define critical angle.
9. Does light travel faster in air or in glass?

Part Two

FIBER-OPTIC COMPONENTS

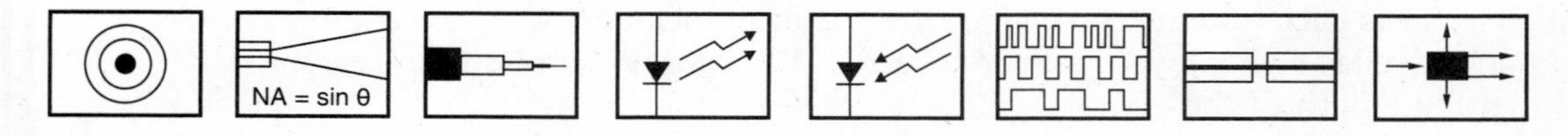

CHAPTER 5

The Optical Fiber

The last chapter showed the characteristics of light propagation most important to your understanding of fiber optics. We saw that the reflection or refraction of light depends on the indices of refraction of the two media and on the angle at which light strikes the interface. The optical fiber works on these principles. Once light begins to reflect down the fiber, it will continue to do so under normal circumstances. The purpose of this chapter is to describe the propagation of light through the various types of optical fibers. The next chapter further examines the properties of fibers.

Keep in mind the distinction between the optical fiber and the fiber-optic cable. The *optical fiber* is the signal-carrying member, similar in function to the metallic conductor in a wire. But the fiber is usually cabled—that is, placed in a protective covering that keeps the fiber safe from environmental and mechanical damage. This chapter deals specifically with the optical fiber itself.

BASIC FIBER CONSTRUCTION

The optical fiber has two concentric layers called the *core* and the *cladding*. The inner core is the light-carrying part. The surrounding cladding provides the difference in refractive index that allows total internal reflection of light through the core. The index of the cladding is less than 1% lower than that of the core. Typical values, for example, are a core index of 1.47 and a cladding index of 1.46. Fiber manufacturers must carefully control this difference to obtain desired fiber characteristics.

Fibers have an additional coating around the cladding. The coating, which is usually one or more layers of polymer, protects the core and cladding from shocks that might affect their optical or physical properties. The coating has no optical properties affecting the propagation of light within the fiber. This coating, then, is a shock absorber.

Figure 5–1 shows the idea of light traveling through a fiber. Light injected into the fiber and striking the core-to-cladding interface at greater than the critical angle reflects back into the core. Since the angles of incidence and reflection are equal, the reflected light will again be reflected. The light will continue zigzagging down the length of the fiber.

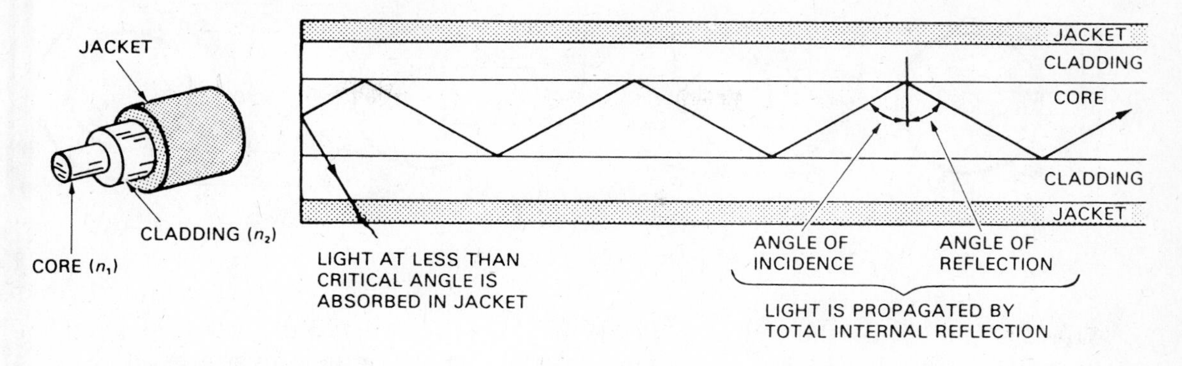

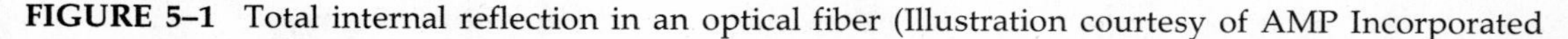
FIGURE 5–1 Total internal reflection in an optical fiber (Illustration courtesy of AMP Incorporated

Light, however, striking the interface at less than the critical angle passes into the cladding, where it is lost over distance. The cladding is usually inefficient as a light carrier, and light in the cladding becomes attenuated fairly rapidly.

Notice also, in Figure 5–1, that the light is also refracted as it passes from air into the fiber. Thereafter, its propagation is governed by the indices of the core and cladding and by Snell's law.

Such total internal reflection forms the basis of light propagation through a simple optical fiber. This analysis, however, considers only *meridional rays*—those that pass through the fiber axis each time they are reflected. Other rays, called *skew rays*, travel down the fiber without passing through the axis. The path of a skew ray is typically helical, wrapping around and around the central axis. Fortunately, skew rays are ignored in most fiber-optic analyses.

The specific characteristics of light propagation through a fiber depend on many factors, including

- The size of the fiber
- The composition of the fiber
- The light injected into the fiber

An understanding of the interplay between these properties will clarify many aspects of fiber optics.

Fibers themselves have exceedingly small diameters. Figure 5–2 shows cross sections of the core and cladding diameters of four commonly used fibers. The diameters of the core and cladding are as follows:

Core (μm)	*Cladding* (μm)
8	125
50	125
62.5	125
100	140

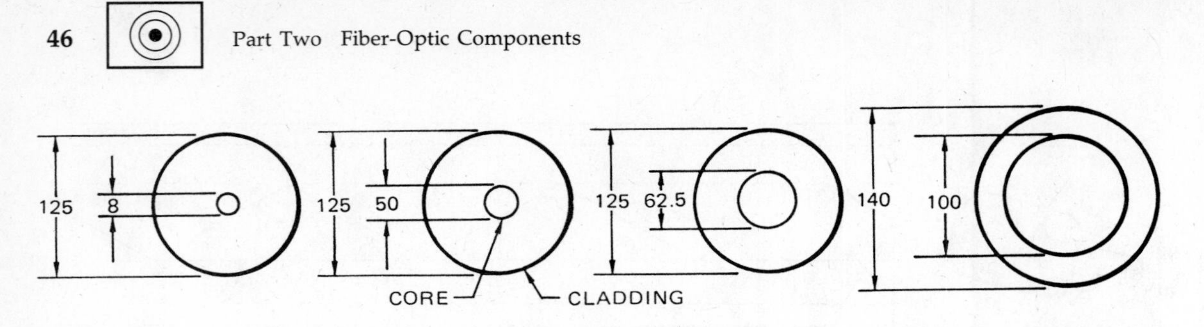

FIGURE 5–2 Typical core and cladding diameters

To realize how small these sizes are, note that human hair has a diameter of about 100 μm. Fiber sizes are usually expressed by first giving the core size, followed by the cladding size: thus, 50/125 means a core diameter of 50 μm and a cladding diameter of 125 μm; 100/140 means a 100-μm core and a 140-μm cladding. Through these small sizes are sent thousands of telephone conversations.

FIBER CLASSIFICATION

Optical fibers are classified in two ways. One way is by their material makeup:

- Glass fibers have a glass core and glass cladding. Since they are by far the most widely used, most discussion in this book centers on glass fibers. The glass used in fibers is ultrapure, ultratransparent silicon dioxide or fused quartz. If seawater were as clear as a fiber, you could see to the bottom of the deepest ocean trench, the 32,177-foot-deep Mariana Trench in the Pacific. Impurities are purposely added to the pure glass to achieve the desired index of refraction. Germanium or phosphorus, for example, increases the index. Boron or fluorine decreases the index. Other impurities not removed when the glass is purified also remain. These, too, affect fiber properties by increasing attenuation by scattering or absorbing light.
- Plastic-clad silica (PCS) fibers have a glass core and plastic cladding. Their performance, though not as good as all-glass fibers, is quite respectable.
- Plastic fibers have a plastic core and plastic cladding. Compared with other fibers, plastic fibers are limited in loss and bandwidth. Their very low cost and easy use, however, make them attractive in applications where high bandwidth or low loss is not a concern. Their electromagnetic immunity and security allow plastic fibers to be beneficially used.

Plastic and PCS fibers do not have the buffer coating surrounding the cladding.

The second way to classify fibers is by the refractive index of the core and the modes that the fiber propagates. Figure 5–3, which depicts the differences in fibers classified this way, shows three important ideas about fibers.

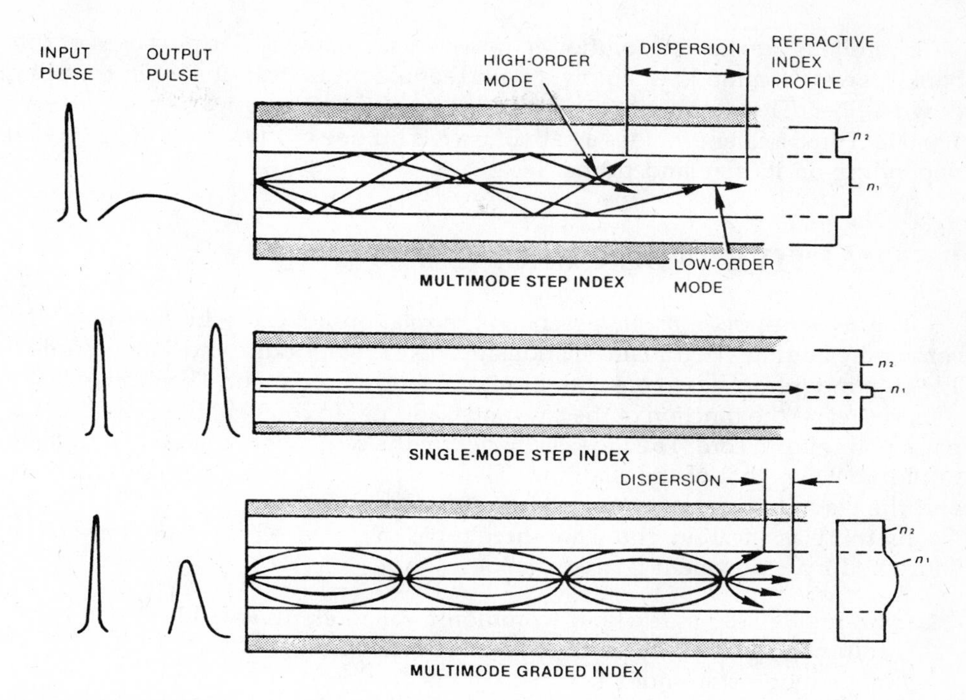

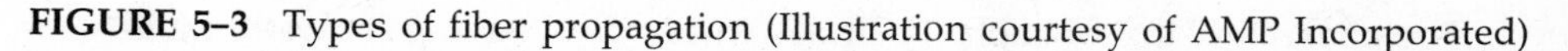
FIGURE 5–3 Types of fiber propagation (Illustration courtesy of AMP Incorporated)

First, it shows the difference between the input pulse injected into a fiber and the output pulse emerging from the fiber. The decrease in height of the pulse shows loss of signal power. The broadening in width limits the fiber's bandwidth or signal-carrying capacity. Second, it shows the path followed by light rays as they travel down the fiber. Third, it shows the relative index of refraction of the core and cladding for each type of fiber. The significance of these ideas will become apparent as we examine each type of fiber.

MODES

Mode is a mathematical and physical concept describing the propagation of electromagnetic waves through media. In its mathematical form, mode theory derives from Maxwell's equations. James Clerk Maxwell, a Scottish physicist in the last century, first gave mathematical expression to the relationship between electric and magnetic energy. He showed that they were both a single form of electromagnetic energy, not two different forms as was then believed. His equations also showed that the propagation of this energy followed strict rules. Maxwell's equations form the basis of electromagnetic theory.

A mode is an allowed solution to Maxwell's equations. For purposes of this book, however, a mode is simply a path that a light ray can follow in traveling down a fiber. The number of modes supported by a fiber ranges from 1 to over 100,000. Thus, a fiber provides a path of travels for one or thousands of light rays, depending on its size and properties.

REFRACTIVE INDEX PROFILE

The *refractive index profile* describes the relation between the indices of the core and cladding. Two main relationships exist: step index and graded index. The step-index fiber has a core with a uniform index throughout. The profile shows a sharp step at the junction of the core and cladding. In contrast, the graded index has a nonuniform core. The index is highest at the center and gradually decreases until it matches that of the cladding. There is no sharp break between the core and the cladding.

By this classification, there are three types of fibers (whose names are often shortened by a prevalent characteristic):

1. Multimode step-index fiber (commonly called step-index fiber)
2. Multimode graded-index fiber (graded-index fiber)
3. Single-mode step-index fiber (single-mode fiber)

The characteristics of each type have important bearing on its suitability for particular applications. As we progress through this chapter and the next, the importance of each type will become apparent.

STEP-INDEX FIBER

The multimode step-index fiber is the simplest type. It has a core diameter from 100 to 970 μm, and it includes glass, PCS, and plastic constructions. As such, the step-index fiber is the most wide ranging, although not the most efficient in having high bandwidth and low losses.

Since light reflects at different angles for different paths (or modes), the path lengths of different modes are different. Thus, different rays take a shorter or longer time to travel the length of the fiber. The ray that goes straight down the center of the core without reflecting arrives at the other end first. Other rays arrive later. Thus, light entering the fiber at the same time exits the other end at different times. The light has spread out in time.

This spreading of an optical pulse is called *modal dispersion.* A pulse of light that began as a tightly and precisely defined shape has dispersed—spread over time. Dispersion describes the spreading of light by various mechanisms. Modal dispersion is that type of dispersion that results from the varying path lengths of different modes in a fiber.

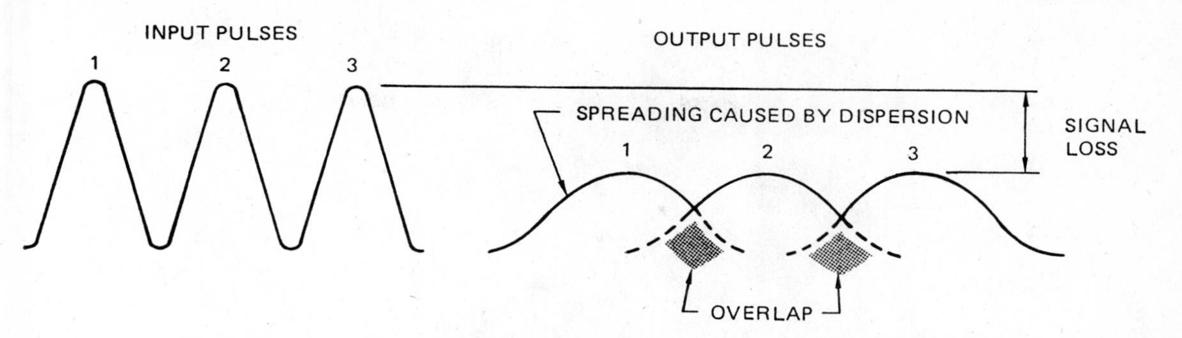

FIGURE 5–4 Pulse spreading

You can image three race cars all traveling the same speed. The first race car follows a straight path—equal to the lowest order mode that does not reflect as it travels. The second car follows the longest path—equal to the highest order mode. While its speed does not change from that of the first car, it must travel continuously back and forth through many curves. The third car follows an intermediate path. If all three cars begin at the same time and travel to a finish line one mile away, they obviously will arrive there at different times. The same holds true for a pulse of light injected into a fiber. Different rays will follow different paths and so arrive at different times.

Typical modal dispersion figures for step-index fibers are 15 to 30 ns/km. This means that when rays of light enter a fiber at the same time, the ray following the longest path will arrive at the other end of a 1-km-long fiber 15 to 30 ns after the ray following the shortest path.

Fifteen to 30 billionths of a second may not seem like much, but dispersion is the main limiting factor on a fiber's bandwidth. Pulse spreading results in a pulse overlapping adjacent pulses, as shown in Figure 5–4. Eventually, the pulses will merge so that one pulse cannot be distinguished from another. The information contained in the pulse is lost. Reducing dispersion increases fiber bandwidth.

GRADED-INDEX FIBER

One way to reduce modal dispersion is to use graded-index fibers. Here the core has numerous concentric layers of glass, somewhat like the annular rings of a tree. Each successive layer outward from the central axis of the core has a lower index of refraction. Figure 5–5 shows the core's structure.

Light, remember, travels faster in a lower index of refraction. So the further the light is from the center axis, the greater its speed. Each layer of the core refracts the light. Instead of being sharply reflected as it is in a step-index fiber, the light is now bent or continually refracted in an almost sinusoidal pattern. Those rays that follow the longest path by traveling near the outside of the core have a faster average

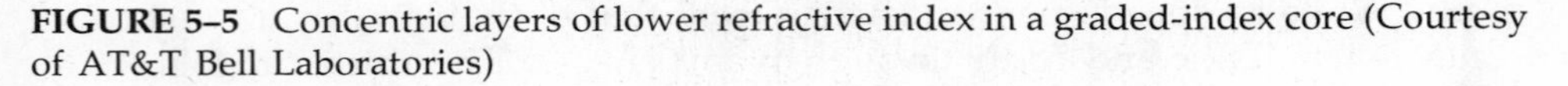

FIGURE 5–5 Concentric layers of lower refractive index in a graded-index core (Courtesy of AT&T Bell Laboratories)

velocity. The light traveling near the center of the core has the slowest average velocity. As a result, all rays tend to reach the end of the fiber at the same time. The graded index reduces modal dispersion to 1 ns/km or less.

Popular graded-index fibers have core diameters of 50, 62.5, or 85 μm and a cladding diameter of 125 μm. The fiber is popular in applications requiring a wide bandwidth, especially telecommunications, local area networks, computers, and similar uses. The 62.5/125-μm fiber is most popular and widely used.

SINGLE-MODE FIBER

Another way to reduce modal dispersion is to reduce the core's diameter until the fiber propagates only one mode efficiently. The single-mode fiber has an exceedingly small core diameter of only 5 to 10 μm. Standard cladding diameter is 125 μm. This cladding diameter was chosen for three reasons:

1. The cladding must be about 10 times thicker than the core in a single-mode fiber. For a fiber with an 8-μm core, the cladding should be at least 80 μm.
2. It is the same size as graded-index fibers, which promotes size standardization.
3. It promotes easy handling because it makes the fiber less fragile and because the diameter is reasonably large so that it can be handled by technicians.

Since this fiber carries only one mode, modal dispersion does not exist.

Single-mode fibers easily have a potential bandwidth of 50 to 100 GHz-km. Present fibers have a bandwidth of several gigahertz and allow transmission of tens of kilometers. As of early 1985, the largest commercially available fiber-optic systems were digital telephone transmission systems operating around 417 Mbps. These systems carried 6048 simultaneous telephone calls over a single-mode fiber a distance of 35 km without a repeater. By the end of 1992, capacities had grown fourfold to 10 Gbps and 130,000 voice channels.

An important aspect of this growth is that the increase results from changing the electronics, not the single-mode fibers, at either end of the system. The capacity of a single-mode system is limited by the capabilities of the electronics, not of the fiber. One advantage of single-mode fibers is that once they are installed, the system's capacity can be increased as newer, higher-capacity transmission electronics becomes available. This capability saves the high cost of installing a new transmission medium to obtain increased performance and allows cost-effective increases in data rates.

The point at which a single-mode fiber propagates only one mode depends on the wavelength of light carried. A wavelength of 820 nm results in multimode operation. As the wavelength is increased, the fiber carries fewer and fewer modes until only one remains. Single-mode operation begins when the wavelength approaches the core diameter. At 1300 nm, for example, the fiber permits only one mode. It becomes a single-mode fiber.

Different fiber designs have a specific wavelength, called the *cutoff* wavelength, above which it carries only one mode. A fiber designed for single-mode operation at 1300 nm has a cutoff wavelength of around 1200 nm.

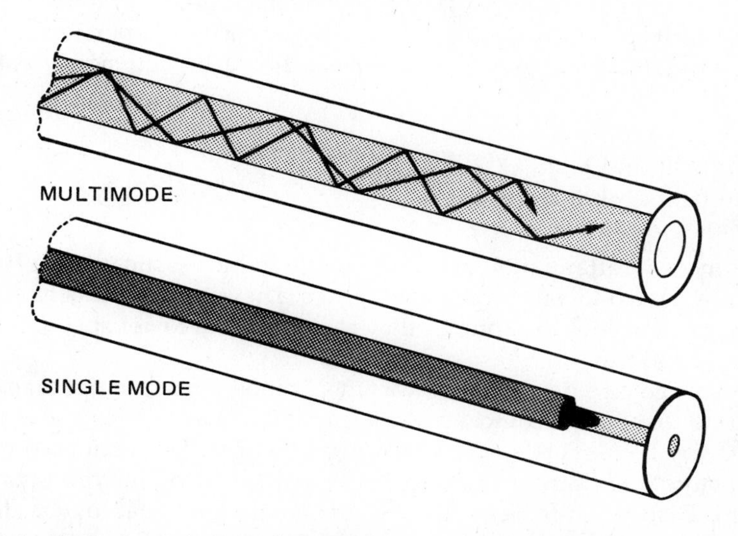

FIGURE 5–6 Optical power in multimode and single-mode fibers (Courtesy of Corning Glass Works)

The operation of a single-mode fiber is slightly more complex than simply a ray traveling down the core. Geometric optics using light rays is not as appropriate for these fibers because it obscures how optical energy is distributed within the fiber. Some of the optical energy of the mode travels in the cladding, as shown in Figure 5–6. Therefore, the diameter of the light appearing at the end of the fiber is larger than the core diameter. *Mode field diameter* is the term used to define this diameter of optical energy. Although optical energy is confined to the core in a multimode fiber, it is not so confined in a single-mode fiber. It is usually more important to know the mode field diameter than the core diameter.

The difference in propagation of light in a single-mode fiber points to another difference between single-mode and multimode fibers. Because optical energy in a single-mode fiber travels in the cladding as well as in the core, the cladding must be a more efficient carrier of energy. In a multimode fiber, the light transmission characteristics of the cladding are basically unimportant. Indeed, because cladding modes are not desirable, a cladding with inefficient transmission characteristics can be tolerated. This situation does not hold for a single-mode fiber.

DISPERSION-SHIFTED FIBERS

Not all single-mode fibers use a step profile. Some use more complex designs in order to optimize the fiber to operate at a certain wavelength. For example, a step-index fiber has zero material dispersion at 1300 nm. Zero dispersion—discussed further in the next chapter—is important because it is at this wavelength that a fiber has its greatest information-carrying capacity. Dispersion is about five times higher at 1550 nm. Attenuation levels, however, are significantly lower at 1550 nm:

- 0.35 to 0.50 dB/km at 1300 nm
- 0.20 to 0.30 dB/km at 1550 nm

The differences in attenuation and dispersion characteristics at the two wavelengths represent a tradeoff to system designers. They can operate at higher speeds and shorter distances at 1300 nm, or they can operate at lower speeds and longer distances at 1550 nm.

The advanced core designs of newer single-mode fibers attempt to make lower loss and low dispersion coincide at the same wavelength, so the system can operate at very high speeds over very long distances. Dispersion-shifted fibers have a structure that changes the zero-dispersion wavelength, typically from 1300 to 1550 nm. Dispersion-flattened fibers have a structure that lowers dispersion over a wide range of wavelengths.

SHORT-WAVELENGTH SINGLE-MODE FIBERS

A single-mode fiber can be constructed with a shorter cutoff wavelength. Some fibers have been designed with a cutoff wavelength of 570 nm for operation at 633 nm (which is visible red light). The core is quite small, less than 4 μm. Another fiber has a cutoff wavelength of 1000 nm, a recommended operating frequency of 1060 nm, and a core diameter of under 6 μm. These fibers are intended for specialized telecommunications, computer, and sensor applications. They are not a replacement for standard long-distance telecommunication single-mode fibers operating at the longer wavelengths of 1300 and 1550 nm. For one thing, their attenuation levels are several times higher—up to 10 dB/km for the 633-nm fiber—making them unsuited for longer distances.

PLASTIC FIBERS

Although most of the discussion in this book involves glass fibers, plastic fibers should not be overlooked. Compared to glass fibers, they have modest performance. A high-end application is 50 Mbps over 100 meters. Yet this performance is good compared to twisted-pair copper cable. Plastic fibers have a relatively large core and very thin cladding. Typical sizes are 480/500, 735/750, and 980/1000 μm, although tolerances in plastic fiber are much greater than in glass fibers. A plastic fiber with a nominal diameter of 480 μm for the core and 500 μm for the cladding may actually have diameters that vary as much as 15 μm in either direction.

Plastic fibers have several distinct features that make them attractive in cost-sensitive applications.

Plastic fibers—and related components like sources, detectors, and connectors—are considerably less expensive than their glass-fiber counterparts.

Plastic fibers use red light in the 660-nm range. Visible light aids diagnostics and troubleshooting since the presence of light in the system is easy to see. In addition, the optical power and output patterns do not pose the safety problem associated with infrared laser light in glass-fiber systems.

Plastic fibers are quite rugged, with a tight bend radius and the ability to withstand abuse. Their EMI immunity makes them attractive in noisy environments.

Finally, the fibers are simple for a technician to work with. Connectors are quickly and easily applied, typically in a minute or less.

Because of the low cost, good performance, and ruggedness of plastic fibers, they find use in such applications as automobiles, music systems, and other consumer electronics. The Japanese, for example, have created standards to use plastic fibers in home electronic systems to interconnect such equipment as digital audio tape and compact disk players.

HOW MANY MODES ARE THERE?

We have seen that the number of modes supported by a fiber in part determines its information-carrying capacity. Modal dispersion, which causes pulse spreading and overlapping, limits the data rate that a fiber can support. We have also seen that dispersion depends on wavelength and core diameter.

The *V* number, or normalized frequency, is a fiber parameter that takes into account the core diameter, wavelength propagated, and fiber NA (a fiber property described in the next chapter):

$$V = \frac{2\pi d}{\lambda}\,(\mathrm{NA})$$

From the *V* number, the number of modes in a fiber can be calculated.

For a simple step-index fiber, the number of modes can be approximated by

$$N = \frac{V^2}{2}$$

For a graded-index fiber, the number of modes can be approximated by

$$N = \frac{V^2}{4}$$

The equations demonstrate that the number of modes is determined by core diameter, fiber NA, and the wavelength propagated. The number of modes in a graded-index fiber is about half that of a step-index fiber having the same diameter and NA. A fiber with a 50-μm core supports over 1000 modes.

When the *V* number of a step-index fiber becomes 2.405, the fiber supports a single mode. The *V* number can be decreased by decreasing the core diameter, by increasing the operating wavelength, or by decreasing the NA. Thus, single-mode operation in a fiber can be obtained by suitably adjusting these characteristics.

Figure 5–7 shows the number of modes contained in three different common fiber sizes operating at two different wavelengths. In the same fiber, the longer wavelength of 1300 nm travels in half as many modes as the shorter 850-nm wavelength. Similarly, decreasing the core diameter also significantly reduces the number of modes.

FIBER COMPARISONS

Table 5–1 shows typical characteristics of various fibers. The meaning of terms such as NA are explained in the next chapter. For now, though, you can see that the performance and physical properties cover a broad range. We use the term ''performance'' broadly: Better performance means higher bandwidth, higher

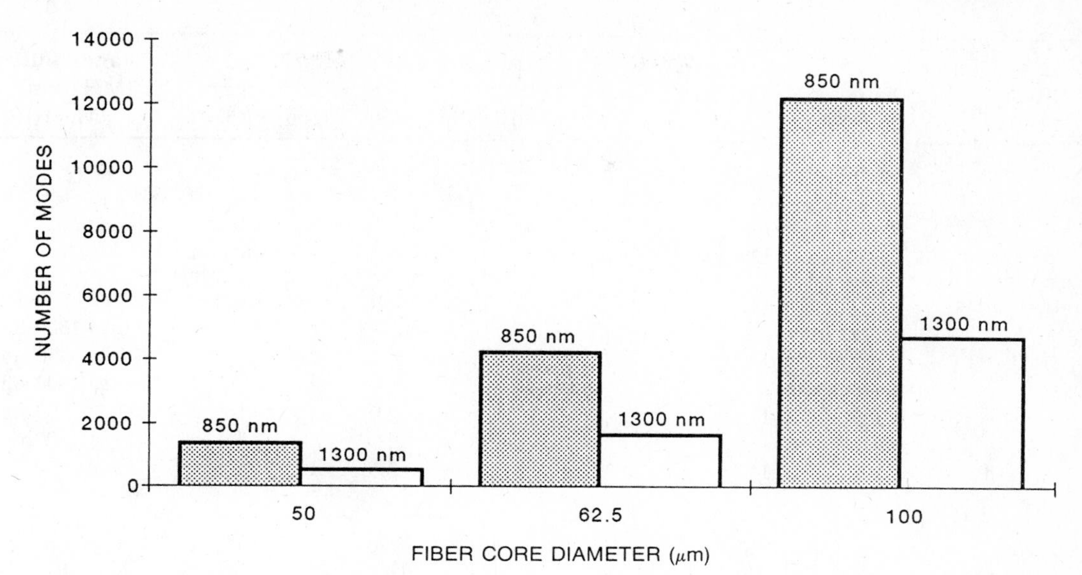

FIGURE 5–7 Number of modes in various fibers at two different wavelengths

information-carrying capacity, and lower losses. Other measures of performance, such as safety or low cost, would favor other types of fibers. The table also suggests some generalizations about loss and bandwidth:

- Fiber performance from lowest to highest is as follows:
  —Plastic
  —PCS
  —Step-index glass
  —Graded index
  —Single mode
- A smaller core usually means better performance.
- Glass fibers perform better than plastic ones.

Remember, though, that such generalizations suffer the fault of all generalizations: They do not tell the whole story. A fiber-optic cable must be matched to the requirements of the application. A system needing to transmit only a few thousand bits a second over a couple of meters can use plastic fibers to best advantage. Not only does plastic fiber cost less, but compatible components such as sources, detectors, and connectors also cost less. Using a single-mode fiber in such applications would be like using a Ferrari to go to the corner store. It is common to stress the low losses and high bandwidths of fibers, but not all applications need such performance. All the fiber types have their uses.

Fiber Type	Core Diameter (μm)[1]	Cladding Diameter (μm)	NA	Attenuation (dB/km) (Max)					Bandwidth (MHz/km) (Max)
				650	790	850	1300	1550	
Single Mode	3.7	80 or 125		10					
	5.0	85 or 125				2.3			5000 @ 850 nm
	9.3	125	0.13				0.4	0.3	6 ps/km[2]
	8.1	125	0.17				0.5	0.25	
Graded Index	50	125	0.20			2.4	0.6	0.5	600 @ 850 nm
									1500 @ 1300 nm
	62.5	125	0.275			3.0	0.7	0.3	200 @ 850 nm
									1000 @ 1300 nm
	85	125	0.26			2.8	0.7	0.4	200 @ 850 nm
									400 @ 1300 nm
	100	140	0.29			3.5	1.5	0.9	300 @ 850 nm
									500 @ 1300 nm
Step Index	200	380	0.27			6.0			6 @ 850 nm
	300	440	0.27			6.0			6 @ 850 nm
PCS	200	350	0.30		10				20 @ 790 nm
Plastic	485	500	0.5	240					5 @ 680 nm[3]
	735	750	0.5	230					
	980	1000	0.5	220					

1. Mode field diameter given for single-mode fiber; actual core diameter is less.
2. Dispersion per nanometer of source width.
3. Plastic fibers typically are used at distances under 100 m, with data rates up to 50 Mbit/s.

TABLE 5–1 Typical fiber characteristics

SUMMARY

- The three types of fibers are step index, graded index, and single mode.
- Dispersion is one factor limiting fiber performance. One motive of fiber design is to reduce dispersion by grading the index or by using a single-mode design.
- Core diameter gives a rough estimate of fiber performance: the smaller the core, the higher the bandwidth and the lower the loss.
- Fibers come in a variety of performances to suit different application needs.
- Mode field diameter is the diameter of optical energy carried in a single-mode fiber.
- Single-mode fibers use both step-index and more complex profiles.

? REVIEW QUESTIONS

1. What are the two main parts of an optical fiber?
2. What are the three types of fibers according to material composition?
3. Light travels down a step-index fiber by what principle?
4. What usually happens to the performance of a fiber if the core diameter is reduced? If it is enlarged?
5. What is the path called that is followed by optical energy down a fiber?
6. What is the name of a fiber that allows only one path for the light down the fiber?
7. What is the name of a fiber with a core whose refractive index varies?
8. Is the refractive index of the core higher or lower than that of the cladding?
9. (True/False) Modal dispersion is constant at all wavelengths.
10. (True/False) One advantage of the graded-index fiber is that the wavelengths of low loss and minimum dispersion always coincide.

Fiber Characteristics

This chapter examines the characteristics of optical fibers most important to users and designers. The chapter expands on the characteristics discussed in previous chapters and introduces new ones.

DISPERSION

As discussed in the last chapter, *dispersion* is the spreading of a light pulse as it travels down the length of an optical fiber. Dispersion limits the bandwidth or information-carrying capacity of a fiber. The bit rate must be low enough to ensure that pulses do not overlap. A lower bit rate means that the pulses are farther apart and, therefore, that greater dispersion can be tolerated. There are three main types of dispersion:

1. Modal dispersion
2. Material dispersion
3. Waveguide dispersion

MODAL DISPERSION

As shown in the last chapter, modal dispersion occurs only in multimode fibers. It arises because rays follow different paths through the fiber and consequently arrive at the other end of the fiber at different times. Modal dispersion can be reduced in three ways:

1. Use a smaller core diameter, which allows fewer modes. A 100-μm-diameter core allows fewer paths than a 200-μm-diameter core.
2. Use a graded-index fiber so that the light rays that follow longer paths also travel at a faster average velocity and thereby arrive at the other end of the fiber at nearly the same time as rays that follow shorter paths.
3. Use a single-mode fiber, which permits no modal dispersion.

MATERIAL DISPERSION

Different wavelengths (colors) also travel at different velocities through a fiber, even in the same mode. Earlier, we saw that the index of refraction is equal to

$$n = \frac{c}{v}$$

where c is the speed of light in a vacuum and v is the speed of the same wavelength in the material.

Each wavelength, though, travels at a different speed through a material, so the value of v in the equation changes for each wavelength. Thus, index of refraction changes according to the wavelength. Dispersion from this phenomenon is called *material dispersion* since it arises from material properties of the fiber. The amount of dispersion depends on two factors:

1. The range of light wavelengths injected into the fiber. A source does not normally emit a single wavelength; it emits several. This range of wavelengths, expressed in nanometers, is the spectral width of the source. An LED has a much wider spectral width than a laser—about 35 nm for an LED and 2 to 3 nm for a laser.
2. The center operating wavelength of the source. Around 850 nm, longer (''reddish'') wavelengths travel faster than shorter (''bluish'') ones. An 860-nm wave travels through glass faster than an 850-nm wave. At 1550 nm, however, the situation is reversed: The shorter wavelengths travel faster than the longer ones, a 1560-nm wave travels slower than a 1540-nm wave. At some point, the crossover must occur where the bluish and reddish wavelengths travel at the same speed. This crossover occurs around 1300 nm, the zero-dispersion wavelength. Figure 6–1 shows this idea. The length of the arrows suggest the speed of the wavelength; hence, a longer arrow is traveling faster and arriving earlier.

Figure 6–2 is a graph showing dispersion for a typical single-mode fiber. Dispersion is zero at 1300 nm. At wavelengths below 1300 nm, dispersion is negative, so wavelengths trail or arrive later. Above 1300 nm, wavelengths lead or arrive faster.

Material dispersion is of greater concern in single-mode systems. In a multimode system, modal dispersion is usually significant enough that material dispersion is not a factor. In many cases, designers do not even concern themselves with modal dispersion. Speeds are too low, or distances are too short.

The 820-to-850-nm region of the spectrum is used as a transmission wavelength for many fiber-optic systems. In this region, material dispersion can be approximated as being roughly equal to 0.1 ns/nm of spectral width.

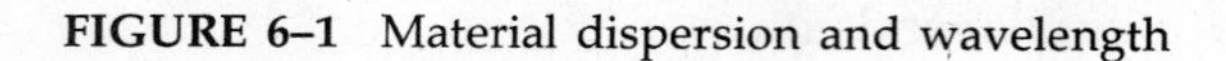

FIGURE 6–1 Material dispersion and wavelength

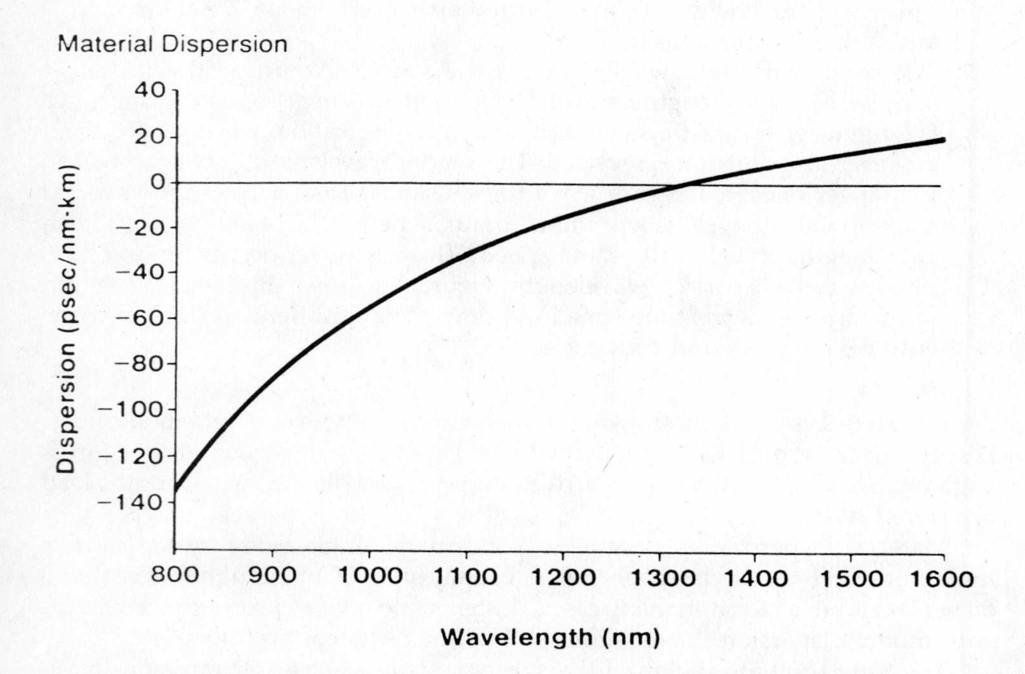

FIGURE 6–2 Material dispersion and the zero-dispersion wavelength (Courtesy of Corning Glass Works)

WAVEGUIDE DISPERSION

Waveguide dispersion, most significant in a single-mode fiber, occurs because optical energy travels in both the core and cladding, which have slightly different refractive indices. The energy travels at slightly different velocities in the core and cladding because of the slightly different refractive indices of the materials. Altering the internal structure of the fiber allows waveguide dispersion to be substantially changed, thus changing the specified overall dispersion of the fiber, which is one goal of the advanced single-mode designs discussed in the last chapter.

BANDWIDTH AND DISPERSION

Many fiber and cable manufacturers do not specify dispersion for their multimode offerings. Instead, they specify a figure of merit called the bandwidth-length product or simply bandwidth, given in megahertz-kilometers. A bandwidth of 400 MHz-km means that a 400-MHz signal can be transmitted for 1 km. It also means that the product of the frequency and the length must be 400 or less ($BW \times L \leq 400$). In other words, you can send a lower frequency a longer distance or a higher frequency a shorter distance, as shown in Figure 6–3.

Single-mode fibers, on the other hand, are specified by dispersion. This dispersion is expressed in picoseconds per kilometer per nanometer of source spectral width (ps/km/nm). In other words, for any given single-mode fiber, dispersion is most affected by the source's spectral width: the wider the source

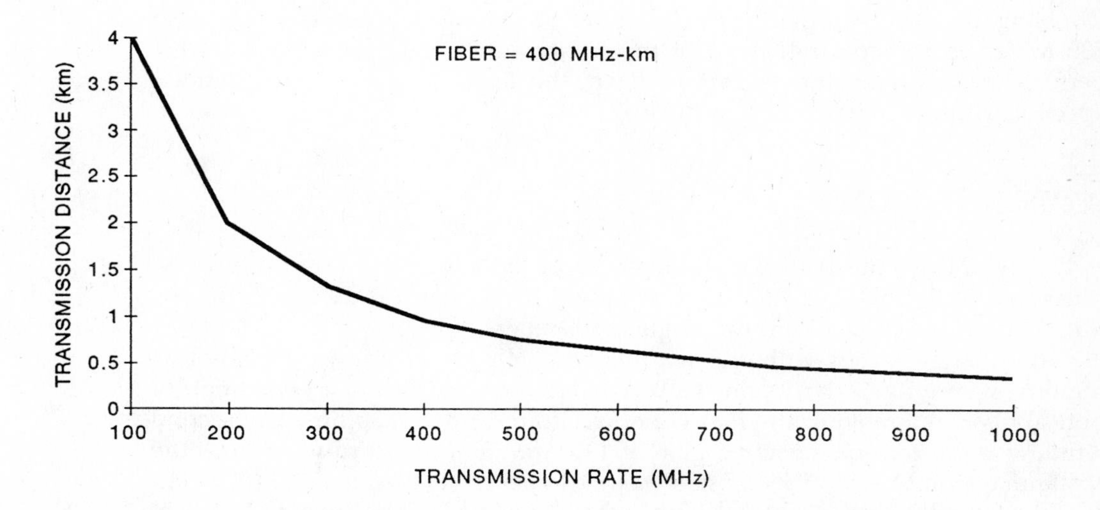

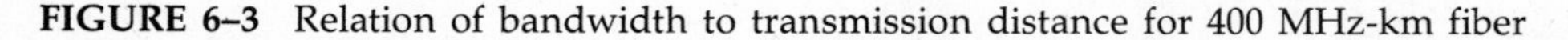
FIGURE 6–3 Relation of bandwidth to transmission distance for 400 MHz-km fiber

width (the greater the range of wavelengths injected into the fiber), the greater the dispersion. Although the conversion of single-mode dispersion to bandwidth is complex, a rough approximation can be gained from the following equation:

$$BW = \frac{0.187}{(\text{Disp})\,(\text{SW})\,(L)}$$

where

Disp = dispersion at the operating wavelength, in seconds per nanometer per kilometer
SW = the spectral width (rms) of the source, in nanometers
L = fiber length, in kilometers

Here is an example using the following:

Dispersion = 3.5 ps/ns/km
Spectral width = 2 nm
Length = 25 km

Substituting these values into the equation yields

$$BW = \frac{0.187}{(3.5 \times 10^{-12}\text{s/nm/km})\,(2\text{ nm})\,(25\text{ km})}$$
$$= 1068\text{ MHz} \quad \text{or, roughly,} \quad 1\text{ GHz}$$

Doubling the source width to 4 nm significantly reduces the bandwidth to around 535 MHz. So the spectral width of the source has a significant effect on the performance of a single-mode fiber. Reducing the dispersion figure or the source's spectral width increases the bandwidth.

ATTENUATION

Attenuation is the loss of optical power as light travels through the fiber. Measured in decibels per kilometer, it ranges from over 300 dB/km for plastic fibers to around 0.21 dB/km for single-mode fibers.

Attenuation varies with the wavelength of light. Windows are low-loss regions, where fibers carry light with little attenuation. The first generation of optical fibers operated in the first window, around 820 to 850 nm. The second window is the zero-dispersion region of 1300 nm, and the third window is the 1550-nm region. A typical 50/125 graded-index fiber offers attenuation of 4 dB/km at 850 nm and 2.5 dB/km at 1300 nm, which is about a 30% increase in transmission efficiency.

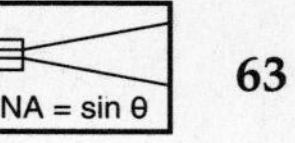

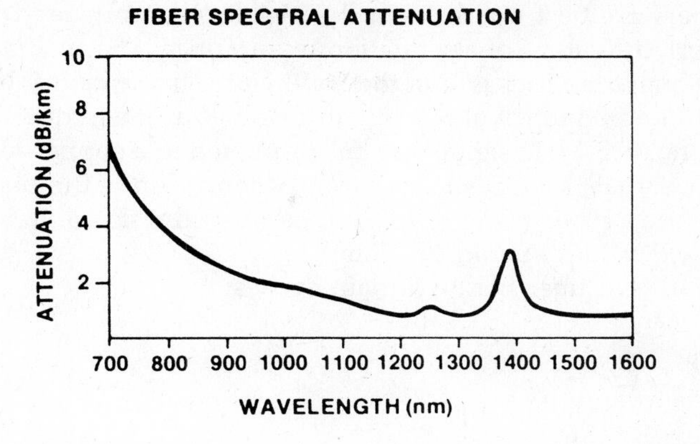

FIGURE 6–4 Attenuation versus wavelength for a multimode fiber (Courtesy of Corning Glass Works)

High-loss regions, where attenuation is very high, occur at 730, 950, 1250, and 1380 nm. One wishes to avoid operating in these regions. Evaluating loss in a fiber must be done with respect to the transmitted wavelength. Figure 6–4 shows a typical attenuation curve for a low-loss multimode fiber. Figure 6–5 does the same for a single-mode fiber; notice the high loss in the mode-transition region, where the fiber shifts from multimode to single-mode operation.

Making the best use of the low-loss properties of the fiber requires that the source emit light in the low-loss regions of the fiber.

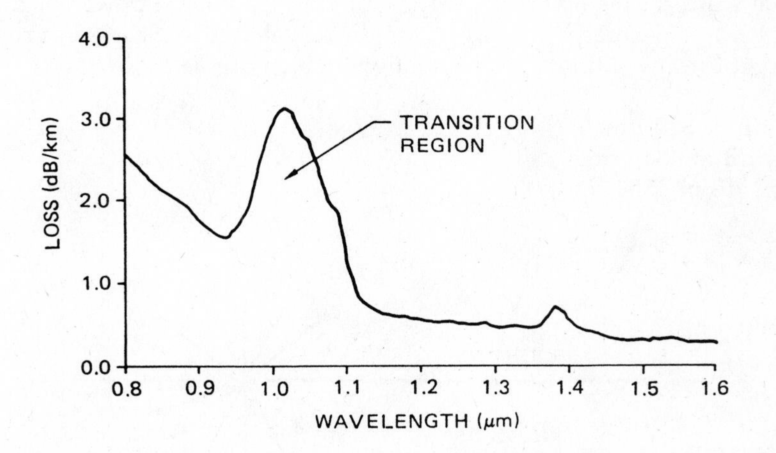

FIGURE 6–5 Attenuation versus wavelength for a single-mode fiber (Courtesy of Corning Glass Works)

Plastic fibers are best operated in the visible-light area around 650 nm.

One important feature of attenuation in an optical fiber is that it is constant at all modulation frequencies within the bandwidth. In copper cables, attenuation increases with the frequency of the signal: The higher the frequency, the greater the attenuation. A 25-MHz signal will be attenuated in a copper cable more than will a 10-MHz signal. As a result, signal frequency limits the distance a signal can be sent before a repeater is needed to regenerate the signal. In an optical fiber, both signals will be attenuated the same.

Attenuation in a fiber has two main causes:

1. Scattering
2. Absorption

SCATTERING

Scattering is the loss of optical energy due to imperfections in the fiber and from the basic structure of the fiber. Scattering does just what the term implies: It scatters the light in all directions (Figure 6–6). The light is no longer directional.

Rayleigh scattering is the same phenomenon that causes a red sky at sunset. The shorter blue wavelengths are scattered and absorbed while the longer red wavelengths suffer less scattering and reach our eyes, so we see a red sunset.

Rayleigh scattering comes from density and compositional variations in a fiber that are natural by-products of manufacturing. Ideally, pure glass has a perfect molecular structure and, therefore, uniform density throughout. In real glass, the density of the glass is not perfectly uniform. The result is scattering.

Since scattering is inversely proportional to the fourth power of the wavelength ($1/\lambda^4$), it decreases rapidly at longer wavelengths. Scattering represents the theoretical lower limits of attenuation, which are as follows:

- 2.5 dB at 820 nm
- 0.24 dB at 1300 nm
- 0.012 dB at 1550 nm

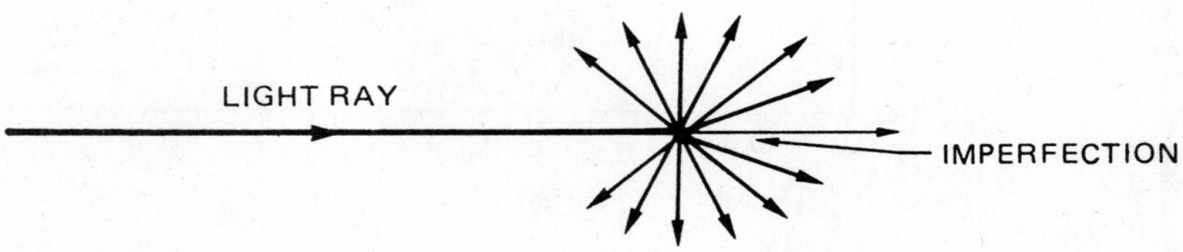

LIGHT RAY

IMPERFECTION

FIGURE 6–6 Scattering

ABSORPTION

Absorption is the process by which impurities in the fiber absorb optical energy and dissipate it as a small amount of heat. The light becomes ''dimmer.'' The high-loss regions of a fiber result from water bands, where hydroxyl molecules significantly absorb light. Other impurities causing absorption include ions of iron, copper, cobalt, vanadium, and chromium. To maintain low losses, manufacturers must hold these ions to less than one part per billion. Fortunately, modern manufacturing techniques, including making fibers in very clean environments, permits control of impurities to the point that absorption is not nearly as significant as it was just a few years ago.

MICROBEND LOSS

Microbend loss is that loss resulting from microbends, which are small variations or ''bumps'' in the core-to-cladding interface. As shown in Figure 6–7, microbends can cause high-order modes to reflect at angles that will not allow further reflection. The light is lost.

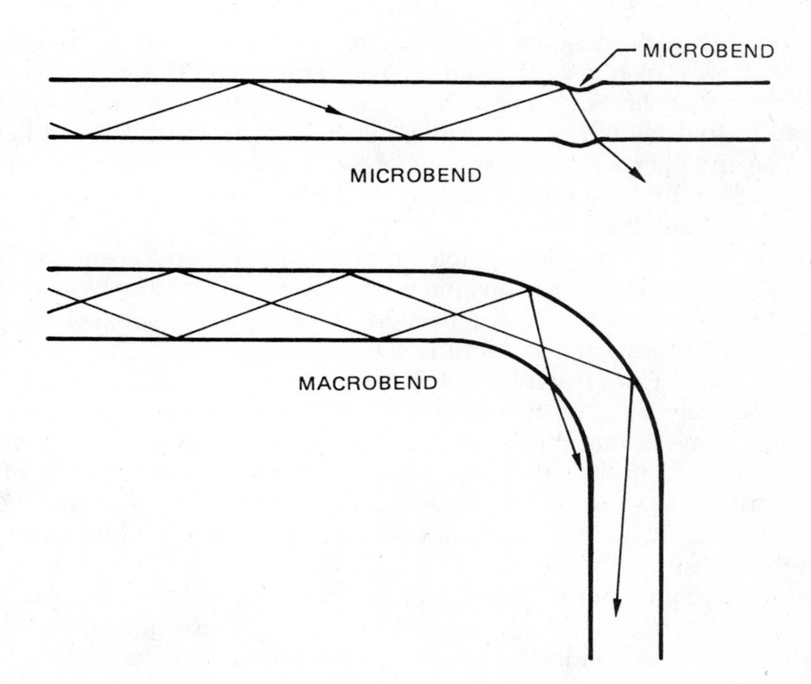

FIGURE 6–7 Loss and bends

Microbends can occur during the manufacture of the fiber, or they can be caused by the cable. Manufacturing and cabling techniques have advanced to minimize microbends and their effects.

EQUILIBRIUM MODE DISTRIBUTION

An important concept about modes in a fiber is that of *equilibrium mode distribution.* While many modes or paths are available to carry light, not all carry the same amount of energy. Nor do all carry light efficiently. Some modes carry no light—meaning that no energy travels along a potential path. What's more, energy can transfer between modes; it can change paths.

In a perfect fiber, the energy in each mode will stay in that mode. But in a real fiber, energy transfer between modes is caused by bends in the fiber, variations in the diameter or refractive index of the core, or other imperfections.

Over distance, light will transfer between modes until it arrives at EMD. At this point, further transfer of energy between modes does not occur under normal circumstances. It can occur under unusual circumstances such as flaws in the fiber, bends in the cable, and other things. At EMD, inefficient modes have lost their optical energy.

Before it reaches EMD, a fiber is said to be *overfilled* or *underfilled.* An overfilled fiber is one in which marginal modes carry optical energy. This energy will be attenuated or lost over a short distance. You can think of this as excess energy, since for many applications it is unimportant. It will be lost over distance. Some light sources, notably LEDs, can overfill a fiber. This means they inject light into modes that the fiber will not carry efficiently. Some of these modes are in the cladding. Others are high-order modes in the core that will not propagate efficiently.

An underfilled fiber is one in which the light injected into the fiber fills only some of the low-order modes available for propagation of optical energy. A laser, for example, because of its narrow, intense beam, might fill only the lowest-order modes—those traveling with few reflections. Over distance, some of this energy will enter higher-order modes until EMD is reached.

Think of a fiber as a water hose. If you try to couple a flow of water too large for the hose, only some of the water will travel through the hose. Some of the water might travel along the outside of the hose (the cladding), but only for a short distance. If, on the other hand, you shoot a very narrow beam of water into the hose, the water at first does not fill the entire hose. Over distance, however, it will. Thus, the hose will reach a steady state or EMD as the water travels through it.

The modal conditions of a short length of fiber depend on the characteristics of the source that injects the light. An LED often overfills the fiber. Over distance, however, the modal conditions become independent of the source.

The distance required to reach EMD varies with the type of fiber. A plastic fiber requires only a few meters or less to reach EMD. A high-quality glass fiber can require tens of kilometers before it reaches EMD.

Equilibrium mode distribution is important to understand for two reasons. First, loss of optical power—attenuation—in an optical fiber depends on modal conditions. In a short length of fiber that has not reached EMD, loss is proportional to length. For a fiber that has reached EMD, loss is proportional to the *square root of length.*

The second reason is the effect that modal conditions has on other conditions in a fiber. Consider the following example as an illustration. Suppose a 1-meter length of fiber is connected to a light source that overfills the fiber. At the end of the fiber we measure the optical energy and find 750 μW of energy. However, this energy includes energy in modes that will be lost by the time the fiber reaches EMD. If, on the other hand, we modify the fiber conditions by wrapping the fiber around a small-diameter mandrel five times, we simulate EMD in this short length of fiber. And now, perhaps, we find only 500 μW of energy emerging from the fiber.

What happened to the other 250 μW? The difference of 250 μW is due to the energy that will be lost before the fiber reaches EMD.

To be able to compare accurately two fibers, two light sources, or two connectors, you must know the conditions under which their manufacturers test them. If one manufacturer uses a fully filled fiber and another uses a fiber under conditions of EMD, the apparent test results will differ dramatically—even if the two fibers are identical. Most fiber-optic test measurements today are performed with fibers at EMD so that comparisons are meaningful.

Another consequence of EMD is the effects it has on two other characteristics of a fiber: numerical aperture and active diameter.

NUMERICAL APERTURE

Numerical aperture (NA) is the ''light-gathering ability'' of a fiber. Only light injected into the fiber at angles greater than the critical angle will be propagated. The *material* NA relates to the refractive indices of the core and cladding:

$$NA = \sqrt{n_1^2 - n_2^2}$$

Notice that NA is dimensionless.

We can also define the angles at which rays will be propagated by the fiber. These angles form a cone, called the *acceptance cone,* that gives the maximum angle of light acceptance. The acceptance cone is related to the NA:

$$\theta = \arcsin(NA)$$
$$NA = \sin \theta$$

where θ (theta) is the half-angle of acceptance. See Figure 6–8.

The NA of a fiber is important because it gives an indication of how the fiber accepts and propagates light. A fiber with a large NA accepts light well; a fiber with a low NA requires highly directional light.

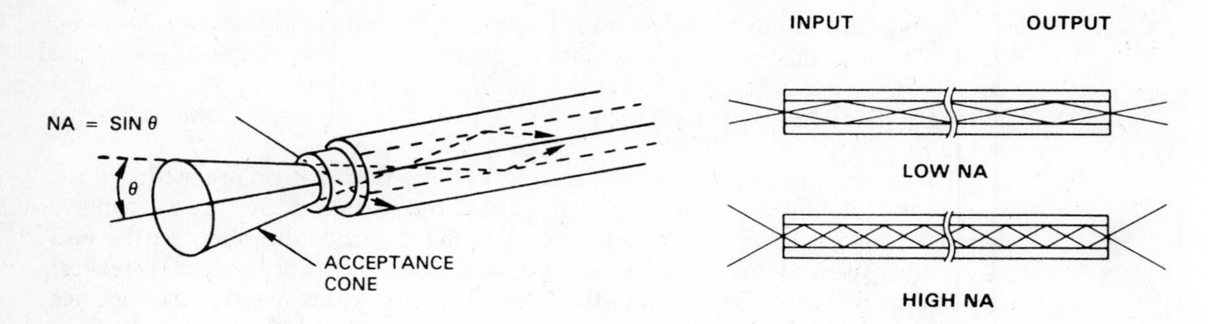

FIGURE 6–8 Numerical aperture (Illustration courtesy of AMP Incorporated)

In general, fibers with a high bandwidth have a lower NA. They thus allow fewer modes. Fewer modes mean less dispersion and, hence, greater bandwidth. NAs range from about 0.50 for plastic fibers to 0.20 for graded-index fibers. A large NA promotes more modal dispersion, since more paths for the rays are provided.

Manufacturers do not normally specify NA for their single-mode fibers (the NA is only about 0.11), because NA is not a critical parameter for the system designer or user. Light in a single-mode fiber is not reflected or refracted, so it does not exit the fiber at angles. Similarly, the fiber does not accept light rays at angles within the NA and propagate them by total internal reflection. As a result, NA, although it can be defined for a single-mode, is not useful as a practical characteristic. Later chapters will show how NA in a multimode fiber is important to system performance and to calculating anticipated performance.

The NA of a fiber changes over distance. High-order modes—those that travel near the critical angle—are often lost. As a graded-index fiber reaches EMD, for example, its NA can be reduced by as much as 50%. This reduction means that the light exiting the fiber does so at angles much less than defined by the acceptance cone. In addition, the spot diameter of light emerging from the fiber can also be reduced. Figure 6–9 shows the diameter of light in the core of a 62.5-μm core under fully filled and EMD conditions. When the fiber is fully filled, the light fills the core. When the fiber reaches EMD, the light diameter is only 50 μm. The NA of the light is similarly reduced.

Sources and detectors also have an NA. The NA of a source defines the angles of the exiting light. The NA of the detector defines the angles of light that will operate the detector. Especially for sources, it is important to match the NA of the source to the NA of the fiber so that all the light emitted by the source is coupled into the fiber and propagated. Mismatches in NA are sources of loss when light is coupled from a lower NA to a higher one.

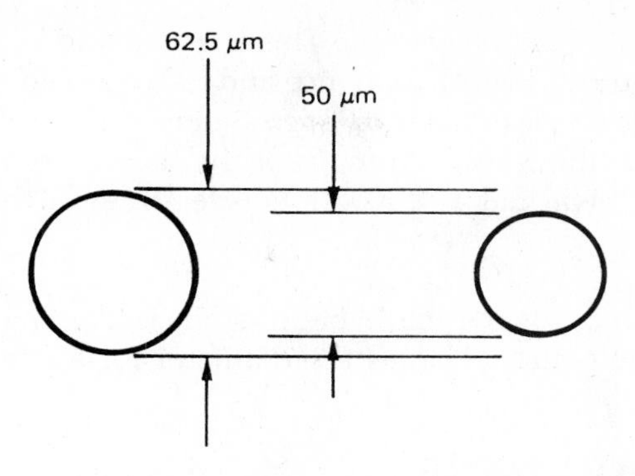

FIGURE 6–9 Active area carrying optical power in a 62.5-μm core under fully filled and EMD conditions (Illustration courtesy of AMP Incorporated)

FIBER STRENGTH

One usually thinks of glass as brittle. Certainly, a pane of glass is not easily bent, let alone rolled up. Yet, a fiber can be looped into tight circles without breaking. Furthermore, fiber can be tied into loose knots. (Pulling the knot tight will break the fiber.)

Tensile strength is the ability of a fiber or wire to be stretched or pulled without breaking. The tensile strength of a fiber exceeds that of a steel filament the same size. Furthermore, a copper wire must have twice the diameter to have the same tensile strength as a fiber.

The main causes of weakness in a fiber are microscopic cracks on the surface and flaws within the fiber. Surface cracks are probably most significant. Surface defects can grow with pulling from the tensile load applied as the fiber is installed and from the tensile load on the fiber during its installed lifetime. Temperature changes, mechanical and chemical damage, and normal aging also promote defects.

Defects can grow, eventually causing the fiber to break. If you have ever cut glass, you understand this phenomenon. To cut glass, you make a shallow scribe across the glass. Given a sharp snap, the glass splits along the scribe. The same effect holds true for optical fibers. A flaw acts like the line scribed in the pane of glass. As the fiber is pulled, the flaw grows into the fiber until the fiber breaks.

BEND RADIUS

Even though fibers can be wrapped in circles, they have a minimum bend radius. A sharp bend will snap the glass. Bends have two other effects:

1. They increase attenuation slightly. This effect should be intuitively clear. The bends change the angles of incidence and reflection enough that some high-order modes are lost (similar to microbends.)
2. Bends decrease the tensile strength of the fiber. If pull is exerted across a bend, the fiber will fail at a lower tensile strength than if no bend were present.

As a rule of thumb, the minimum bend radius is five times the cable diameter for an *unstressed* cable and 10 times the diameter for a *stressed* cable.

NUCLEAR HARDNESS

Nuclear hardness refers to the ability of equipment to withstand nuclear effects. The effect of nuclear radiation on a conductor is of great interest to the U.S. military (in particular, to protect and maintain its command, control, and communication [C^3] system), to the nuclear power industry, and to other areas of high-radiation risk. Fibers are nonconductive and do not build up static charges when exposed to radiation. Fibers will not short out if their jackets are melted by the high heat of a nuclear emergency.

Fibers do suffer an increase in attenuation when subjected to high levels of continuous nuclear radiation. Nuclear radiation increases the light absorption by the fiber's impurities. The increase in attenuation depends on the amount of the dosage and the intensity of the dosage. A high-intensity burst of 3700 rad in 3 ns increases attenuation to peaks of thousands of decibels per kilometer. This excess attenuation decreases to under 10 dB/km in less than 10 s and to fewer than 5 dB/km in less than 100 s. Thus, glass fibers are able to transmit information after exposure to high-radiation weapons bursts within minutes after the radiation has cleared.

An electromagnetic pulse (EMP) is another concern associated with nuclear weapons, although its effects are more closely related to severe EMI than to nuclear radiation. An EMP results from a nuclear explosion. Two or three nuclear devices detonated over the United States at an altitude of several hundred kilometers could damage or destroy every piece of unprotected electronic equipment in the country.

Gamma rays, produced in the first few nanoseconds after detonation, travel until they collide with electrons of air molecules in the upper atmosphere. These electrons, scattered and accelerated by the collisions, are deflected by the earth's magnetic field to produce a transverse electric current. This current sets up EMPs that radiate downward. Any metal conductors will pick up these pulses and conduct them. A 1-megaton warhead can produce peak fields of 50 kV/m with an instantaneous peak power of around 6 MW/m^2. Such levels are far beyond the ability of equipment to withstand. In short, the entire power and communications grid of the country could be wiped out by an electromagnetic pulse.

SUMMARY

- Dispersion is a general term for phenomena that cause light to spread as it travels through a fiber.
- The three types of dispersion are modal, material, and waveguide.
- Dispersion limits bandwidth.
- Dispersion in multimode fibers includes modal and material dispersion.
- Dispersion in single-mode fibers includes material and waveguide dispersion. Material dispersion is the most important.
- Attenuation is the loss of signal power.
- Attenuation varies with the frequency of light.
- Attenuation does not vary with the signal rate in an optical fiber.
- Numerical aperture is the light-gathering ability of a fiber. It defines the angles at which a fiber will accept and propagate light.
- Fibers have a higher tensile strength than do wires of comparable size.

? REVIEW QUESTIONS

1. Name three types of dispersion.
2. Which type of dispersion does not exist in a single-mode fiber?
3. Is the information-carrying capacity of a multimode fiber characterized by dispersion or by bandwidth?
4. Is the information-carrying capacity of a single-mode fiber characterized by dispersion or by bandwidth?
5. If a multimode fiber has a bandwidth of 250 MHz-km, how far can a 750-MHz signal be transmitted?
6. Name the two main mechanisms of attenuation in an optical fiber.
7. Is attenuation lower in a fiber at 850 nm, 1300 nm, or 1550 nm? Why?
8. For any particular single-mode fiber, what specific characteristic limits its data rate?
9. Does an optical fiber recover from a short, intense exposure from nuclear radiation in seconds, minutes, hours, or days?
10. (True/False) The material NA of a fiber always indicates the modal conditions of light traveling within the fiber.

CHAPTER 7

Fiber-Optic Cables

Most often, an optical fiber must be packaged before it is used. Packaging involves cabling the fiber. *Cabling* is an outer protective structure surrounding one or more fibers. It is analogous to the insulation or other materials surrounding a copper wire. Cabling protects copper and fibers environmentally and mechanically from being damaged or degraded in performance. (The additional protection against electrical shock, shorts, and the possibility of fire important to copper is not an issue with dielectric fibers.) The purpose of this chapter is to describe some typical cable structures.

Like their copper counterparts, fiber-optic cables come in a great variety of configurations (Figure 7–1). Important considerations in any cable are tensile strength, ruggedness, durability, flexibility, environmental resistance, temperature extremes, and even appearance.

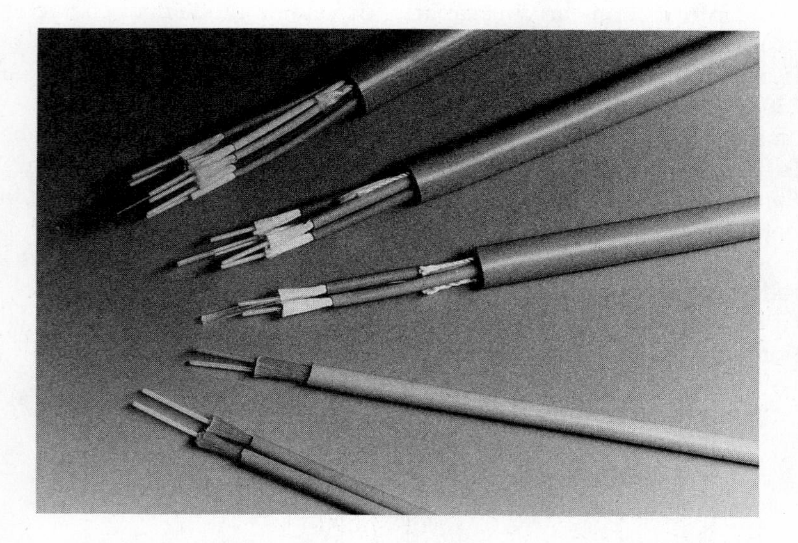

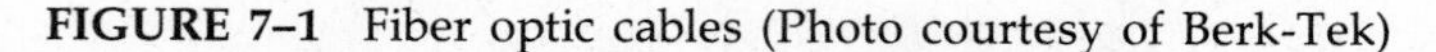

FIGURE 7–1 Fiber optic cables (Photo courtesy of Berk-Tek)

Evaluation of these considerations depends on the application. An outside plant telephone cable must endure extreme hardship. It must withstand extremes of heat and cold, ice deposits that cause it to sag on a pole, high winds that buffet it, and rodents that chew on it underground. It obviously must be more rugged than a cable connecting equipment within the controlled environment of a telephone switching office. Similarly, a cable running under an office carpet, where people will walk on it and chairs will roll over it, has different requirements than a cable running within the walls of the office.

MAIN PARTS OF A FIBER-OPTIC CABLE

Figure 7–2 shows the main parts of a simple single-fiber cable. Even though cables come in many varieties, most have the following parts in common:

- Optical fiber
- Buffer
- Strength member
- Jacket

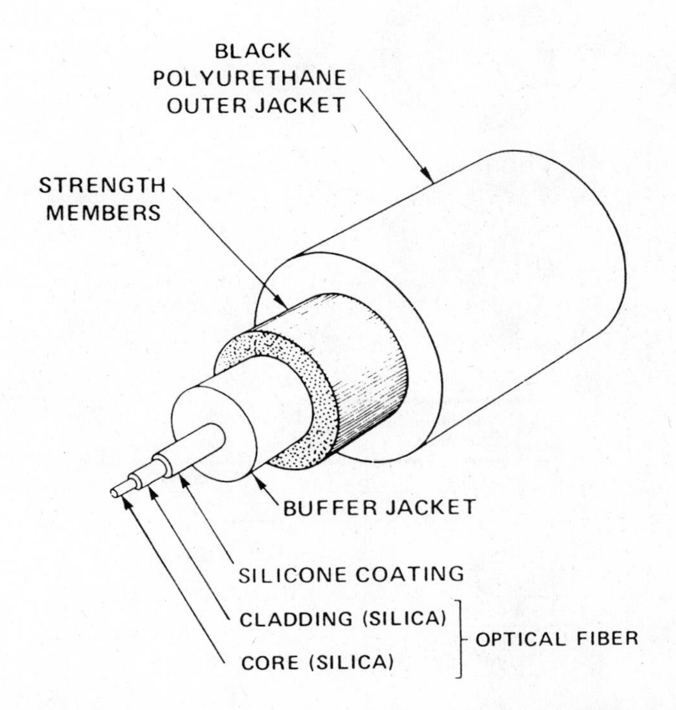

FIGURE 7–2 Parts of a fiber-optic cable (Courtesy of Hewlett-Packard)

Since we have already looked extensively at optical fibers, we discuss only the buffer, strength member, and jacket here.

BUFFERS

The simplest buffer, discussed in Chapter 5, is the plastic coating applied to the cladding. This buffer, which is part of the fiber, is applied by the fiber manufacturer. An additional buffer is added by the cable manufacturer. (Most cable manufacturers and vendors do not make their own fibers.)

The cable buffer is one of two types: loose buffer or tight buffer. Figure 7–3 shows the two constructions and summarizes the tradeoffs involved with each.

The *loose buffer* uses a hard plastic tube having an inside diameter several times that of the fiber. One or more fibers lie within the buffer tube. The tube isolates the fiber from the rest of the cable and the mechanical forces acting on it. The

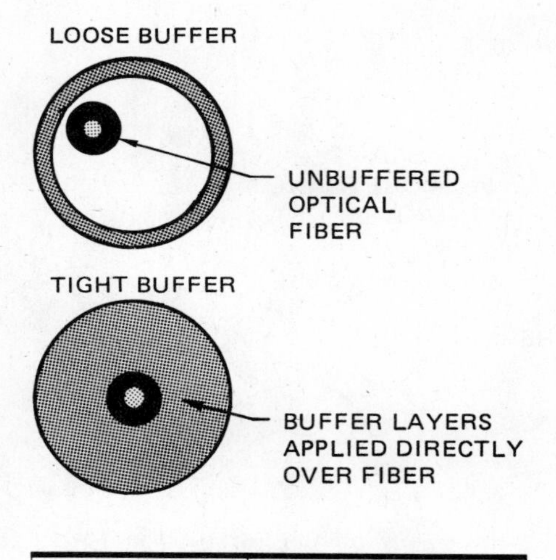

Cable Parameter	Cable Structure	
	Loose Tube	Tight Buffer
Bend Radius	Larger	Smaller
Diameter	Larger	Smaller
Tensile Strength, Installation	Higher	Lower
Impact Resistance	Lower	Higher
Crush Resistance	Lower	Higher
Attenuation Change At Low Temperatures	Lower	Higher

FIGURE 7–3 Loose and tight buffers (Courtesy of Belden Electronic Wire and Cable)

buffer becomes the load-bearing member. As the cable expands and shrinks with changes in temperature, it does not affect the fiber as much. A fiber has a lower temperature coefficient than most cable elements, meaning that it expands and contracts less. Typically, some excess fiber is in the tube; in other words, the fiber in the tube is slightly longer than the tube itself. Thus the cable can easily expand and contract without stressing the fiber.

The *tight buffer* has a plastic directly applied over the fiber coating. This construction provides better crush and impact resistance. It does not, however, protect the fiber as well from the stresses of temperature variations. Because the plastic expands and contracts at a different rate than the fiber, contractions caused by variations in temperature can result in loss-producing microbends.

Another advantage to the tight buffer is that it is more flexible and allows tighter turn radii. This advantage can make tight-tube buffers useful for indoor applications where temperature variations are minimal and the ability to make tight turns inside walls is desired.

STRENGTH MEMBERS

Strength members add mechanical strength to the fiber. During and after installation, the strength members handle the tensile stresses applied to the cable so that the fiber is not damaged. The most common strength members are Kevlar aramid yarn, steel, and fiberglass epoxy rods. Kevlar is most commonly used when individual fibers are placed within their own jackets (as in Figure 7–2). Steel and fiberglass members find use in multifiber cables. Steel offers better strength than fiberglass, but it is sometimes undesirable when one wishes to maintain an all-dielectric cable. Steel, for example, attracts lightning, whereas fiberglass does not.

JACKET

The *jacket*, like wire insulation, provides protection from the effects of abrasion, oil, ozone, acids, alkali, solvents, and so forth. The choice of jacket materials depends on the degree of resistance required for different influences and on cost. Table 7–1 compares the relative properties of various popular jacket materials.

When a cable contains several layers of jacketing and protective material, the outer layers are often called the *sheath*. The jacket becomes the layer directly protecting fibers, and the sheath refers to additional layers. This terminology is especially common in the telephone industry.

	PVC	Low Density Polyethylene	Cellular Polyethylene	High Density Polyethylene	Poly-propylene	Poly-urethane	Nylon	Teflon
Oxidation Resistance	E	E	E	E	E	E	E	O
Heat Resistance	G-E	G	G	E	E	G	E	O
Oil Resistance	F	G	G	G-E	F	E	E	O
Low Temperature Flexibility	P-G	G-E	E	E	P	G	G	O
Weather, Sun Resistance	G-E	E	E	E	E	G	E	O
Ozone Resistance	E	E	E	E	E	E	E	E
Abrasion Resistance	F-G	F-G	F	E	F-G	O	E	E
Electrical Properties	F-G	E	E	E	E	P	P	E
Flame Resistance	E	P	P	P	P	F	P	O
Nuclear Radiation Resistance	G	G	G	G	F	G	F-G	P
Water Resistance	E	E	E	E	E	P-G	P-F	E
Acid Resistance	G-E	G-E	G-E	G-E	E	F	P-F	E
Alkali Resistance	G-E	G-E	G-E	G-E	E	F	E	E
Gasoline, Kerosene, Etc. (Aliphatic Hydrocarbons) **Resistance**	P	P-F	P-F	P-F	P-F	G	G	E
Benzol, Toluol, Etc. (Aromatic Hydrocarbons) **Resistance**	P-F	P	P	P	P-F	P	G	E
Degreaser Solvents (Halogenated Hydrocarbons) **Resistance**	P-F	P	P	P	P	P	G	E
Alcohol Resistance	G-E	E	E	E	E	P	P	E

P = poor F = fair G = good E = excellent O = outstanding
These ratings are based on average performance of general purpose compounds. Any given property can usually be improved by the use of selective compounding.

TABLE 7–1 Properties of jacket materials (Courtesy of Belden Electronic Wire and Cable)

INDOOR CABLES

Cables for indoor applications include the following:

- Simplex cables
- Duplex cables
- Multifiber cables
- Heavy-, light-, and plenum-duty cables
- Breakout cables

Although these categories overlap, they represent the common ways of referring to fibers. Figure 7–4 shows cross sections of several typical cable types.

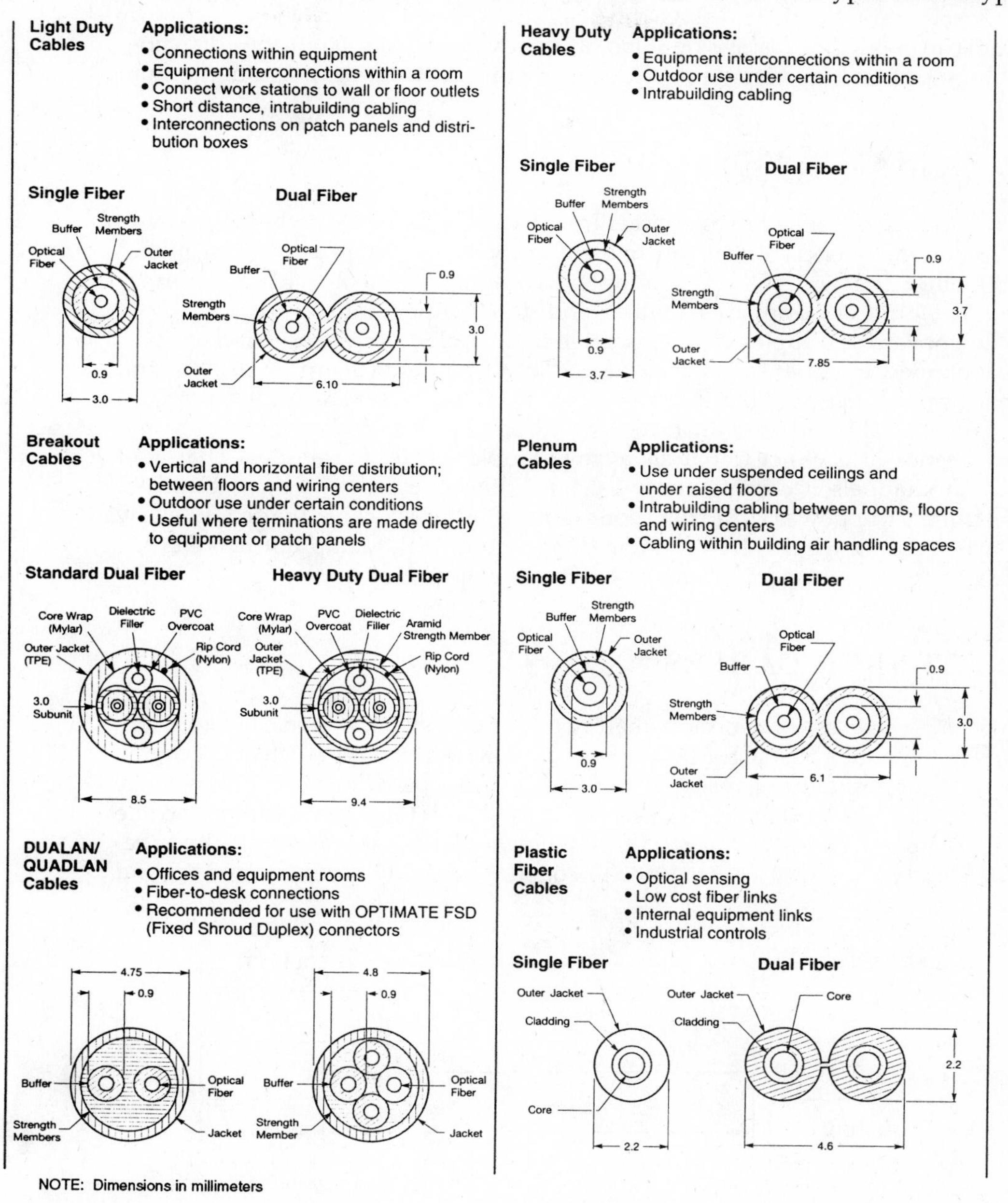

FIGURE 7–4 Indoor cables (Illustration courtesy of AMP Incorporated)

SIMPLEX CABLES

Simplex cables contain a single fiber. ''Simplex'' is a term used in electronics to indicate one-way transmission. Since a fiber carries signals in only one direction, from transmitter to receiver, a simplex cable allows only one-way communication.

DUPLEX CABLES

Duplex cables contain two optical fibers. ''Duplex'' refers to two-way communication. One fiber carries signals in one direction; the other fiber carries signals in the other direction. (Of course, duplex operation is possible with two simplex cables.) Figure 7–5 compares simplex and duplex operations.

In appearance, duplex cables resemble two simplex cables whose jackets have been bonded together, similar to the jacket of common lamp cords. Ripcord constructions, which allow the two cables to be easily separated, are popular.

Duplex cable is used instead of two simplex cables for aesthetics and convenience. It is easier to handle a single duplex cable, there is less chance of the two channels becoming confused, and the appearance is more pleasing. Remember, the power cord from your lamp is a duplex cable that could easily be two separate wires. Does a single duplex cord in the lamp not make better sense? The same reasoning prevails with fiber-optic cables.

MULTIFIBER CABLES

Multifiber cables contain more than two fibers. They allow signals to be distributed throughout a building. Fibers are usually used in pairs, with each fiber carrying signals in opposite directions. A 10-fiber cable, then, permits five duplex circuits.

Multifiber cables often contain several loose-buffer tubes, each containing one or more fibers. The use of several tubes allows identification of fibers by tube, since both tubes and fibers can be color coded. These tubes are stranded around

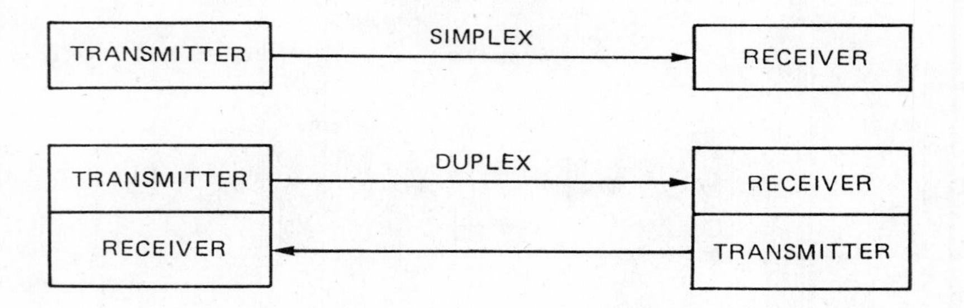

FIGURE 7–5 Simplex and duplex

a central strength member of steel or fiberglass. The stranding provides strain relief for the fibers when the cable is bent.

DUTY SPECIFICATIONS

A cable's construction depends on its application. There are four basic distinctions of application:

- Light duty
- Heavy duty
- Plenum
- Riser

Heavy-duty cables usually have thicker jackets than light-duty cables to allow for rougher handling, especially during installation.

A plenum is the air space between walls, under floor structures, and above drop ceilings. Plenums are popular places to run signal, power, and telephone lines. Unfortunately, plenums are also areas where fires can spread easily through a building.

Certain jacket materials give off noxious fumes when burned. The National Electrical Code (Article 770) requires that cables run through plenums must either be enclosed in fireproof conduits or be insulated and jacketed with low-smoke and fire-retardant materials. Underwriters Laboratories (UL) specifies a procedure called the UL 910 Steiner Tunnel test that rates the flammability of cables. Plenum-rated cables must pass this test and are rated by UL as type OFNP (optical fiber nonconductive plenum) cables.

A riser cable is one that runs vertically between floors of a building. Riser cables must be engineered to prevent fire from spreading between floors and must pass the UL 1666 flame test. They are rated OFNR (optical fiber nonconductive riser) by UL.

BREAKOUT CABLES

Breakout cables have several individual simplex cables inside an outer jacket. The breakout cables shown in Figure 7–4 use two dielectric fillers to keep the cables positioned, while a Mylar wrap surrounds the cables/fillers. The outer jacket includes a ripcord to make its removal fast and easy. The point of the breakout cable is to allow the cable subunits inside to be exposed easily to whatever length is needed. Breakout cables are typically available with two or four fibers, although larger cables also find use.

OUTDOOR CABLES

Cables used outside a building must withstand harsher environmental conditions than most indoor cables. Outdoor cables are used in applications such as the following:

- *Overhead:* Cables are strung from telephone poles.
- *Direct burial:* Cables are placed directly in a trench dug in the ground and covered.
- *Indirect burial:* This is similar to direct burial, but the cable is inside a duct or conduit.
- *Submarine:* The cable is underwater, including transoceanic applications.

Such cables must obviously be rugged and durable, since they will be exposed to various extremes. Most cables have additional protective sheaths. For example, a layer of steel armoring protects against rodents that might chew through plastic jackets and into the fiber. Other constructions use a gel compound to fill the cables and eliminate air within the cable. Loose-tube buffers, for example, are so filled to prevent water from seeping into the cable, where it will freeze, expand, and damage the cable. The fibers ''float'' in the gel that will not freeze and damage the fiber.

Most outdoor cables contain many fibers. The strength member is usually a large steel or fiberglass rod in the center, although small steel strands in the outer sheath are also used. Most high-fiber-count cables divide the fibers among several buffer tubes. Figure 7–6 shows the cross section of several typical constructions.

Another approach is ribbon cable. Here, 12 parallel fibers are sandwiched between double-sided adhesive polyester tape. Each ribbon can be stacked with others to make a rectangular array. For example, a stack of 12 ribbons creates an array of 144 fibers. This array is placed in a loose tube, which in turn is covered by two layers of polyethylene. Each polyethylene layer contains 14 steel wires serving as strength members. Depending on the application, additional layers, such as steel armoring, cover the polyethylene. The first commercial fiber-optic system installed by AT & T, in 1977, used a ribbon structure. Figure 7-7 shows such ribbon cable.

In cables containing many fibers, not all the fibers are always used at first. Some are kept as spares to replace fibers that fail in the future. Others are saved for future expansion, as the demands for additional capacity dictate. Installing a cable can be expensive. Having extra fibers in place when they are needed saves future installation costs of additional fibers and cables. It pays to think ahead.

Siecor® Standard Loose Tube Cable
8 Fiber - Double Jacket - Steel Tape Armor

Central Member
Overcoated As Required
Steel or Dielectric
Fiber
Loose Tube Buffer
Filled
Interstitial Filling
Kevlar
PE Jacket
Steel Tape Armor
PE Jacket

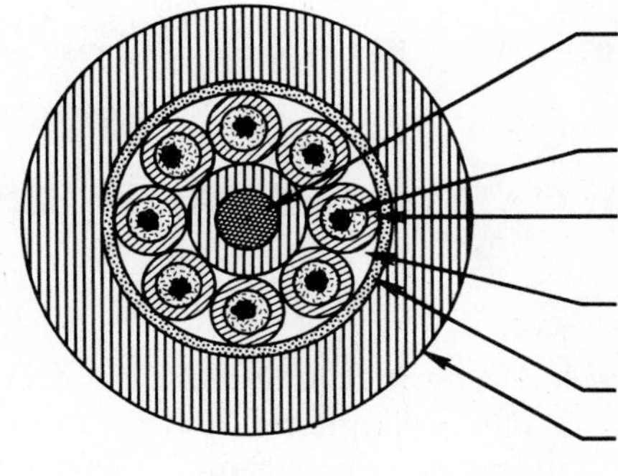

Siecor® Mini-Bundle™ (Loose Tube) Cable
48 Fibers - Underwater Cable

Central Member
Overcoated As Required
Steel or Dielectric
Fiber Bundle
Loose Tube Buffer
Filled
Interstitial Filling
Kevlar
PE Jacket
Steel Tape Armor
PE Jacket
Bedding of Fibrous Fillers
Steel Wire Armoring
Asphalt Jacket

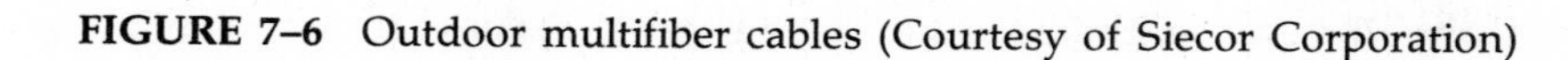

FIGURE 7–6 Outdoor multifiber cables (Courtesy of Siecor Corporation)

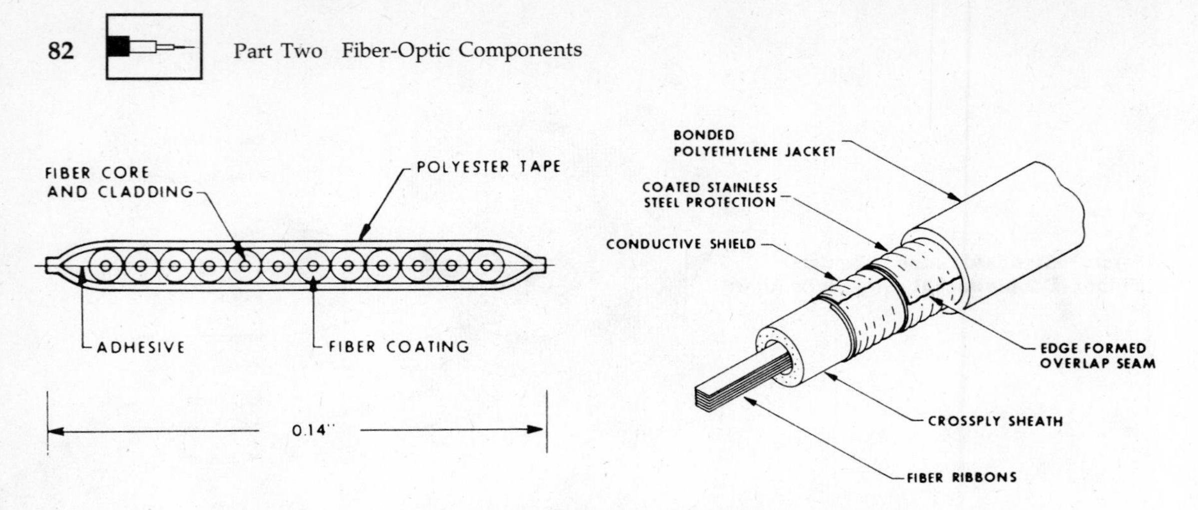

FIGURE 7–7 Ribbon cable (Courtesy of AT&T Bell Laboratories)

ADDITIONAL CABLE CHARACTERISTICS

Lengths

Cables come reeled in various lengths, typically 1 or 2 km, although lengths of 5 or 6 km are available for single-mode fibers. Long lengths are desirable for long-distance applications, since cables must be spliced end to end over the run. Each splice introduces additional loss into the system. Long cable lengths mean fewer splices and less loss.

Color Coding

Fiber coatings or buffer tubes or both are often color coded to make identification of each fiber easier. In a long-distance link, one must ensure that fiber A in the first cable is spliced to fiber A in the second cable, B to B, C to C, and so forth. Color coding simplifies fiber identification.

Loads

Most cable vendors specify the maximum tensile loads that can be applied to the cable. Two loads are usually specified. The *installation load* is the short-term load that the fiber can withstand during the actual process of installation. This load includes the additional load that is exerted by pulling the fiber through ducts or conduits, around corners, and so forth. The maximum specified installation load limits the length of cable that can be installed at one time, given the particular application. Different applications will offer different installation load conditions. One must carefully plan the installation to avoid overstressing the fiber.

The second load specified is the long-term or *operating load*. During its installed life, the cable cannot withstand loads as heavy as it withstood during installation. The specified operating load is therefore less than the installation load. The operating load is also called the *static load*.

Installation and operational loads are specified in pounds or newtons. Allowable loads, of course, depend on cable construction and intended application. A typical specification for a simplex indoor cable is an installation load of 250 lb (1112 N) and an operational load of 10 lb (44 N).

HYBRID CABLES

Fiber-optic cables sometimes also contain copper conductors, such as twisted pairs. Although the conductors can be used for routine communications, they have two other popular uses. One use is to allow installers to communicate with each other during installation of the fiber, especially with long-distance telephone installation. A technician performing a splice in a manhole must often be in communication with a switching office several miles away. The office contains the test equipment, operated by another technician, that tests the quality of the splice. The other use is to power remote electronic equipment such as repeaters.

UNDERSTANDING CABLE SPECIFICATIONS

Table 7–2 reproduces typical specifications for fiber-optic cable. The table lists the specifications most important to the designer planning a fiber-optic system.

Most cable configurations come with various sizes and types of fibers. For example, many fibers have a buffer coating of 250- or 900-μm diameter. This coating allows fibers of 8/125, 50/125, 62.5/125, 85/125, or 100/140 μm to be used. Each of these fibers can further be offered with various attenuations and bandwidths to satisfy the needs of a particular application. In addition, a cable using a loose-tube buffer can hold one or several fibers. None of these factors significantly influences cable construction. The same construction can accommodate all these differences easily.

As fiber-optic technology became widespread, serious debate evolved over which multimode fiber was best suited to different applications. For example, 62.5/125- 85/125-, and 100/140-μm fibers were all proposed for premises wiring and local area networks. The debate centered on the technical and costs merits of each fiber: attenuation, bandwidth, NA, ease and cost of coupling light into the fiber, and so forth.

The ''winner'' of these debates was the 62.5/125-μm fiber, which is the specified or preferred fiber in nearly all applications involving premises wiring, LANs, computer interconnections, and similar uses.

Single-mode fibers are still the preferred choice for long-distance, high-speed applications, while both 50/125- and 100/140-μm are still popular in many applications.

Fiber Size (μm)	Description	Atten. Max. 850 nm (dB/km)	Atten. Max. 1300 nm dB/km)	Bandwidth Min. 850 nm (MHz-km)	Bandwidth Min. 1300 nm (MHz-km)	Dom. Nom. (nm)	Cable Weight (kg/km)	Operating Temp. (°C)	Tensile Load Install. (N)	Bend Radius Min. @ Install. (cm)	Crush Resistance (N/cm)	Flame Rating
	Light Duty Single	3.5	2.0	400	400	3.0	7.5	−40/+70	598	4.5	200	OFNR
	Heavy Duty Single	3.5	2.0	400	400	3.7	12.0	−40/+70	797	5.6	200	OFNR
	Light Duty Dual	3.5	2.0	400	400	3.0×6.0	15.0	−40/+70	1196	4.5	200	OFNR
50/125	Heavy Duty Dual	3.5	2.0	400	400	3.7×7.8	27.0	−40/+70	1595	4.6	200	OFNR
	Plenum Grade Single	3.5	2.0	400	400	3.0	7.5	−40/+70	598	4.5	200	OFNP
	Plenum Grade Dual	3.5	2.0	400	400	3.0×6.0	15.0	−40/+70	1196	4.5	200	OFNP
	DUALAN	3.5	2.0	400	400	4.75	17.0	−40/+70	1595	7.0	200	OFNR
	Light Duty Single	3.5	1.5	160	500	3.0	7.5	−40/+70	598	4.5	200	OFNR
	Heavy Duty Single	3.5	1.5	160	500	3.7	12.0	−40/+70	979	4.5	200	OFNR
	Light Duty Dual	3.5	1.5	160	500	3.0×6.0	15.0	−40/+70	1196	4.5	200	OFNR
	Heavy Duty Dual	3.5	1.5	160	500	3.7×7.8	27.0	−40/+70	1595	5.6	200	OFNP
	Plenum Grade Single	3.5	1.5	160	500	3.0	7.5	−40/+70	598	4.5	200	OFNP
62.5/125	Plenum Grade Dual	3.5	1.5	160	500	3.0×6.0	15.0	−40/+70	1196	4.5	200	OFNP
	DUALAN	3.5	1.5	160	500	4.75	17.0	−40/+70	1595	7.0	200	OFNR
	DUALAN Plenum	3.5	1.5	160	500	4.75	17.0	−40/+70	1595	7.0	200	OFNP
	DUALAN Plenum	3.5	1.5	160	500	4.75	17.0	−40/+70	1595	7.0	200	OFNP
	QUADLAN	3.5	1.5	160	500	4.75	17.0	−40/+70	1595	7.0	200	OFNR
	QUADLAN Plenum	3.5	1.5	160	500	4.75	18.0	−40/+70	1253	7.0	200	OFNP
	Light Duty Single	5.0	4.0	100	200	3.0	7.5	−40/+70	598	4.5	200	OFNR
	Heavy Duty Single	5.0	4.0	100	200	3.7	12.0	−40/+70	797	5.6	200	OFNR
100/140	Light Duty Dual	5.0	4.0	100	200	3.0×6.0	15.0	−40/+70	1196	4.5	200	OFNR
	Heavy Duty Dual	5.0	4.0	100	200	3.7×7.8	27.0	−40/+70	1595	5.6	200	OFNR
	Plenum Grade Single	5.0	4.0	100	200	3.0	7.5	−40/+70	598	4.5	200	OFNR
	Plenum Grade Dual	5.0	4.0	100	200	3.0×6.0	15.0	−40/+70	1196	4.5	200	OFNR
	DUALAN Plenum	5.0	4.0	100	200	4.75	17.0	−40/+70	1595	7.0	200	OFNP
	Light Duty Single	—	1.0	—	—	3.0	7.5	−40/+70	598	4.5	200	OFNR
Singlemode	DUALAN	—	1.0	—	—	4.75	17.0	−40/+70	1595	7.0	200	OFNR
	DUALAN Plenum	—	1.0	—	—	4.75	17.0	−40/+70	1595	7.0	200	OFNP

All cable jackets are PVC.

50/125 Cables:
NA: .20±.02

65.5/125 Cables:
NA: 275±0.15

100/140 Cables:
NA:.29±.02

Ordering Information:
Standard Lengths of 1.0 and 2.0 km
Non-Standard Lengths are available.

TABLE 7–2 Typical cable specifications (Courtesy of AMP Incorporated)

SUMMARY

- Cabling is the packaging of optical fibers for their intended application.
- The main parts of the cable are the buffer, strength member(s), and jacket.
- Cables are available for most application environments.

? REVIEW QUESTIONS

1. Name three materials commonly used as strength members.
2. Name two types of cable buffers.
3. Which is greater: installation load or operating load?
4. A main difference between an indoor cable and an outdoor cable is
   A. Fiber bandwidth
   B. Ruggedness and durability
   C. Number of fibers
   D. Attenuation
5. Describe the difference between simplex and duplex cables.
6. Name two uses for copper wires used in a fiber-optic cable.

Sources

At each end of a fiber-optic link is a transducer, which is simply a device for converting energy from one form to another. The source is an electro-optic transducer—that is, it converts the electrical signal to an optical signal. The detector at the other end is the optoelectronic transducer. It converts optical energy to electrical energy.

The source is either a light-emitting diode (LED) or a laser diode. Both are small semiconductor chips, about the size of a grain of sand, that emit light when current is passed through them. To help you understand the operation of LEDs and lasers, as well as the photodetectors described in Chapter 9, we must first review some of the fundamentals of materials in general and semiconductors in particular.

SOME ATOMIC MATTERS

An atom consists of an inner nucleus around which swirls a cloud of electrons. The electrons circle the nucleus in shells. As shown in Figure 8–1, each shell has a maximum number of electrons in it. The inner shell, K, has a maximum of 2 electrons, L a maximum of 8, M a maximum of 18, and N a maximum of 32. More important for our purposes here, however, is that the outer shell always has a maximum of 8 electrons. This outer shell is called the *valence shell* or *valence*

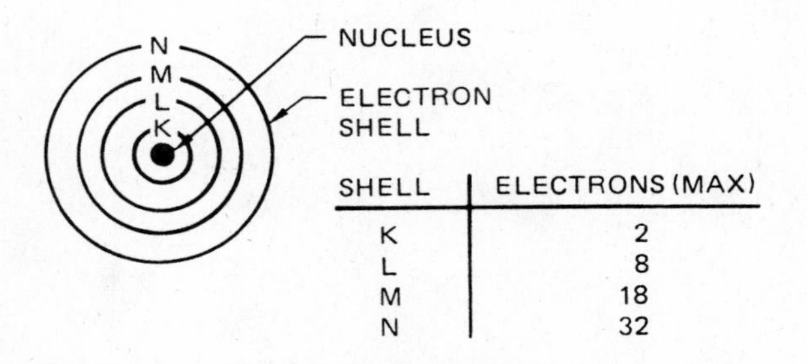

SHELL	ELECTRONS (MAX)
K	2
L	8
M	18
N	32

FIGURE 8–1 Electronic structure of an atom

band. The valence band is involved in the chemical bonds that allow elements and compounds to exist.

For an electron to flow as current, it must be free from the valence band to be able to move about in the molecular structure of the material. An electron so freed from the atom is said to be *in the conduction band*. The electron is a free electron because it no longer is bound to the atom. To be available, then, the electron must move from the valence band to the conduction band. How easily this occurs is what separates good conductors from poor conductors.

An atom with only one electron in the valence shell is a very good conductor. Only a small amount of energy is required to promote the electron to the conduction band as a free electron. Copper, gold, and silver, having only one valence electron, are good conductors.

Iron, cobalt, and platinum are examples of elements with eight valence electrons. As a result, they are poor conductors, because an appreciable amount of external energy is required to separate a bound electron. An atom "wishes" to have a filled outer shell. For copper, it easily releases the outer electron, so the next inner shell is filled. For iron, its shell is filled, and it does not wish to release an electron.

A *semiconductor* is a material whose properties fall between those of good conductors like copper and poor conductors like plastics. Separating an electron from the valence shell requires a medium amount of energy. Silicon, for example, has four electrons in its outer shell.

Energy is involved in the movement of an electron between the valence and conduction bands. For an electron to move from the valence band to the conduction band, external energy must be added. An atom that has lost a valence electron develops forces to attract a free electron and complete its shell. When an electron moves from the conduction band to the valence band, it emits energy. The energy released depends on the difference in energy levels between the two bands involved.

Figure 8–2 shows the two energy bands or levels of interest for our understanding of sources and detectors. For an electron to move from one band to another, energy must be either emitted or absorbed. The energy required is the difference between the energy levels of the two bands: $E_1 - E_2$. This difference

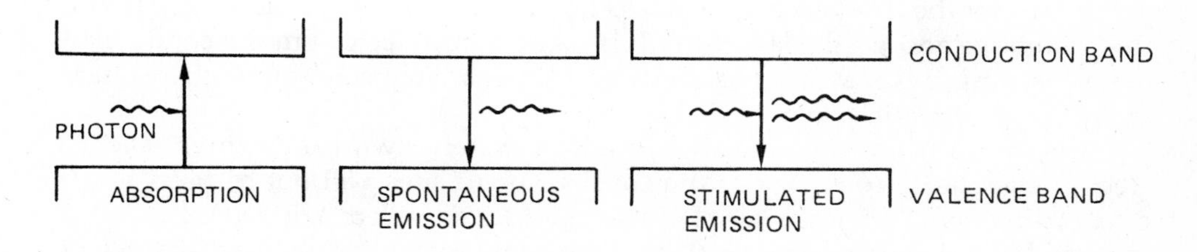

FIGURE 8–2 Absorption and emission

 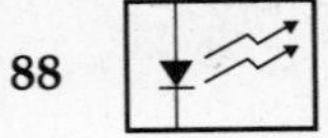

is the *band gap.* If the band gap is proper—that is, if the difference between the band gap is in a given range—the energy is optical energy, or light. The wavelength of the energy is equal to

$$\lambda = \frac{hc}{|E_1 - E_2|}$$

where h is Planck's constant (6.625×10^{-34} J-s), c is the velocity of light, and E_1 and E_2 are the energy levels.

In the spontaneous emission shown in Figure 8–2, an electron randomly moves to the lower energy level. An LED works by spontaneous emission. In stimulated emission, a photon of light causes, or stimulates, an electron to change energy levels. A laser operates on the principle of stimulated emission.

In absorption, external energy, such as supplied by a photon of incident light, supplies the energy to allow the electron to move to the higher energy band.

SEMICONDUCTOR PN JUNCTION

The semiconductor pn junction is the basic structure used in the electro-optic devices for fiber optics. Lasers, LEDs, and photodiodes all use the pn junction, as do other semiconductor devices such as diodes and transistors. We will first describe the basic operation of the junction, and then we will describe the operation of LEDs and lasers. Photodiodes are discussed in greater detail in Chapter 9.

As we saw earlier, a silicon atom has four valence electrons. It is these electrons that form the bonds that hold the atoms together in the crystalline structure of the element. For silicon, these bonds are covalent bonds in which the atoms share their electrons in the bond, as shown in Figure 8–3. Here each atom has access to eight valence electrons required for a full shell, four of its own and four from surrounding atoms. All the electrons are taken up in covalent bonds, and none are readily available as free electrons.

The pn junction begins with such a material as silicon. Suppose we add a material with five valence electrons to a silicon crystal. Since only four electrons are needed for the covalent bonds, an electron is left. This electron is a free electron that can move around the structure. It is really a free electron in the conduction band. Because this material has an excess of negatively charged electrons, it is called *n-type* material.

Suppose, on the other hand, we add a material with only three valence electrons to the silicon. One of the covalent bond sites will not be filled by an electron because none is available. This vacancy is called a *hole.* A hole in a semiconductor is a very strange thing because, by definition, it is an absence of something. In appearance, however, it is a charge carrier similar to an electron, but it

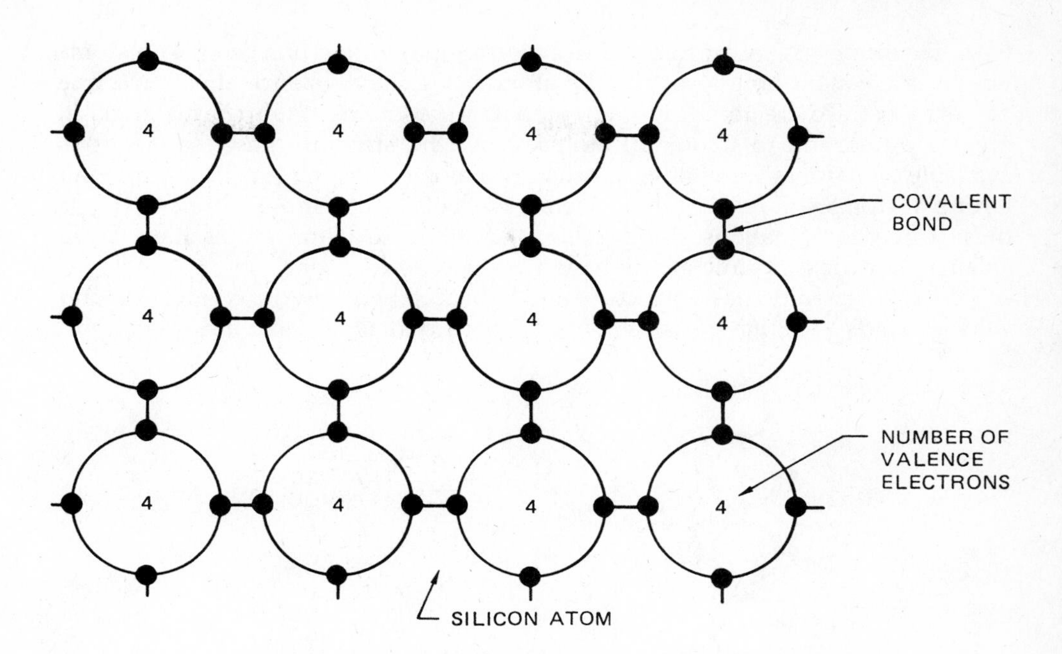

FIGURE 8–3 Covalent bonds in a silicon atom

is positively charged. It appears as a positively charged particle. The material appears to have an excess of positively charged holes, so it is called *p-type* material.

The pn junction of a diode is formed when a semiconductor material is purposely doped with atoms to form p-type and n-type areas separated by a junction. The n-type material has free electrons, and the p-type material has holes. When the p-type and n-type materials are brought together, holes and electrons sweep across the junction and recombine. Recombination means that a free electron ''falls'' into a hole, moving from the conduction band to the valence band. It becomes part of the covalent bonding structure of the atom. Both the hole and the electron disappear as charge carriers.

Because of the recombination of carriers around the pn junction, no carriers exist in the area. A barrier is formed within this depleted region that prevents further migration of electron and hole carriers across the junction unless additional energy is applied to the material.

When a hole and electron recombine, energy is emitted. Depending on the material, this energy may or may not be light. For silicon, the energy is not light. It is heat in the form of vibrations in the crystal structure.

Lasers and LEDs use elements from Groups III and V of the periodic table. These elements have three and five electrons in their valence shells. If we combine an equal number of atoms with three electrons and an equal number of atoms with five electrons, we will have a situation similar to that of silicon. The atoms will form covalent bonds so that the atoms have filled valence shells. No free carriers exist. To create an n-type material, we combine Group V materials in greater proportion to Group III atoms. Now the structure has free electrons available as carriers. Similarly, a structure having more Group III atoms than Group V atoms will result in holes being available as carriers. Figure 8–4 shows these three combinations using gallium arsenide. Gallium atoms have three valence electrons; arsenic atoms have five valence electrons.

Gallium arsenide has a structure and band gap between the conduction and valence bands such that the recombination of carriers results in emission of light.

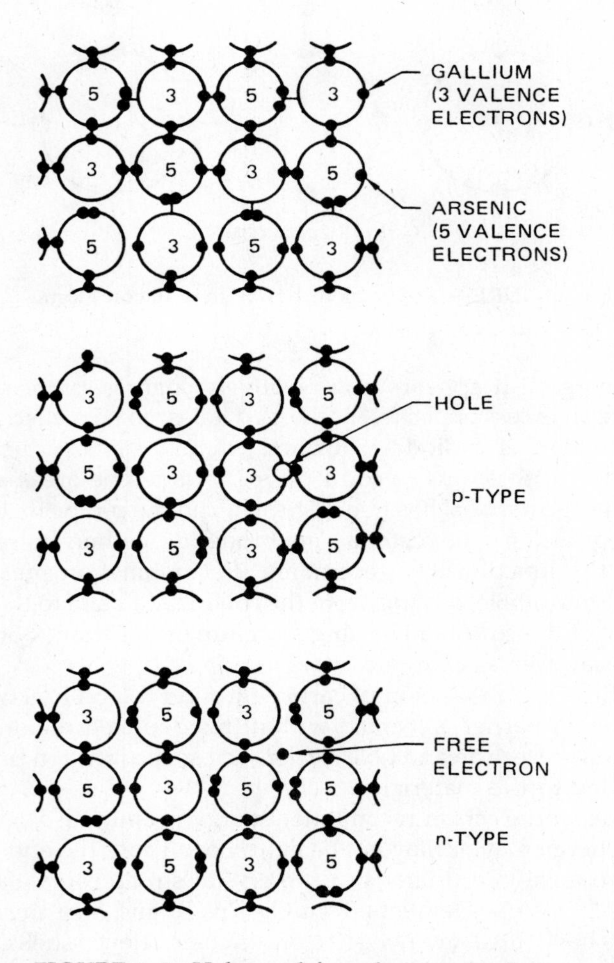

FIGURE 8–4 Holes and free electrons in GaAs

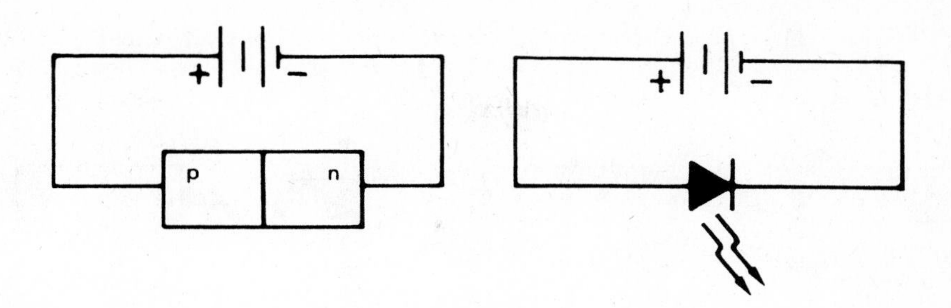

FIGURE 8–5 Light-emitting diode

An LED is a pn semiconductor that emits light when a *forward bias* is applied (the negative battery terminal connects to the n-type material). In a forward bias, electrons are injected into the n-type material and extracted from the p-type material. (Extracting electrons from p-type material is the same as injecting holes into the material.) Figure 8–5 shows the schematic symbol for an LED and its bias arrangement.

The forward bias causes the electrons and holes to move toward one another and cross the *depletion area* of the junction. Electrons and holes combine, emitting light in the process. For the operation to be maintained, current must continually be supplied to maintain an excess of carriers for recombination. When the drive is removed, recombination of carriers around the junction recreates the depletion area and emission ceases.

LEDs

LEDs used in fiber optics are somewhat more complex than the simple device we have just described, although their operation is essentially the same. The complexities arise from the desire to construct a source having characteristics compatible with the needs of a fiber-optic system. Principal among these characteristics are the wavelength and pattern of the emission.

The LED we have described is a *homojunction* device, meaning that the pn junction is formed from a single semiconductor material. Homojunction LEDs are surface emitters giving off light from the edges of the junction as well as its entire planar surface. The result is a low-radiance output whose large pattern is not well suited for use with optical fibers. The problem is that only a very small portion of the light emitted can be coupled into the fiber core.

A *heterojunction* structure solves this problem by confining the carriers to the active area of the chip. A heterojunction is a pn junction formed with materials having similar crystalline structure but different energy levels and refractive indices. The differences confine the carriers and provide a more directional output

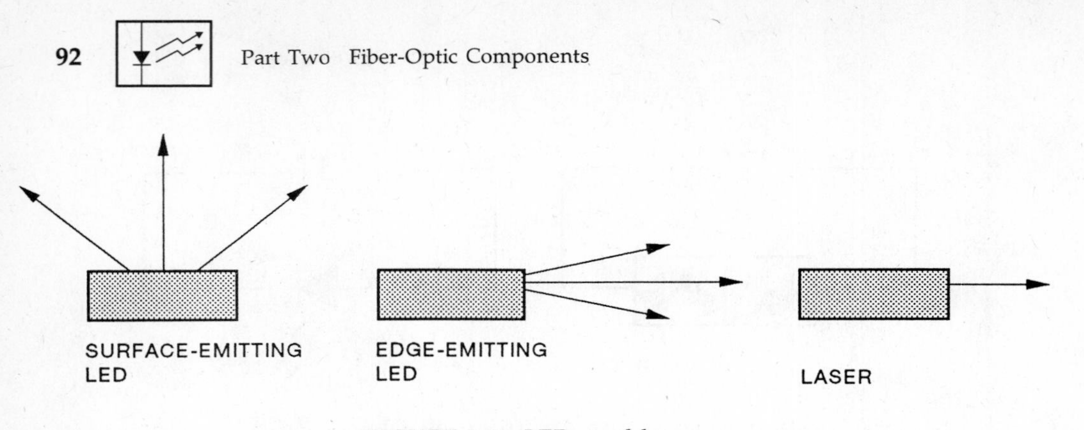

FIGURE 8–6 LEDs and lasers

for the light. The difference in refractive index, for example, can be used to confine and to guide the light in much the same manner that it is confined and guided in an optical fiber. The result is a high-radiance output.

Figure 8–6 shows both a surface-emitting and an edge-emitting LED. The edge-emitting diode uses an active area having a stripe geometry. Because the layers above and below the stripe have different refractive indices, carriers are confined by the waveguide effect produced. (The waveguide effect is the same phenomenon that confines and guides the light in the core of an optical fiber.) The width of the emitting area is controlled by etching an opening in the silicon oxide insulating area and depositing metal in the opening. Current through the active area is restricted to the area below the metal film. The result is a high-radiance elliptical output.

The materials used in the LED influence the wavelengths emitted. LEDs emitting in the first window of 820 to 850 nm are usually gallium aluminum arsenide (GaAIAs). ''Window'' is a term referring to ranges of wavelengths matched to the properties of the optical fiber. One reason this region was the first exploited in fiber optics is that these devices were better understood, more easily manufactured, more reliable, and less costly. In short, the 820-nm technology was mature, as the 1300-nm technology now is. The movement from 1300 to 1550 nm depends in part on improved technology for sources.

Long-wavelength devices for use at 1300 nm are made of gallium indium arsenide phosphate (GaInAsP) and other combinations of Group III and Group V materials.

LASERS

Laser is an acronym for *l*ight *a*mplification by the *s*timulated *e*mission of *r*adiation. Lasers (Figure 8–6) provide stimulated emission rather than the simpler spontaneous emission of LEDs. The main difference between an LED and a laser is that the laser has an optical cavity required for lasing. This cavity, called a *Fabry-Perot cavity,* is formed by cleaving the opposite end of the chip to form highly parallel, reflective mirrorlike finishes.

At low drive currents, the laser acts like an LED and emits light spontaneously. As the current increases, it reaches the threshold level, above which lasing action begins. A laser relies on high current density (many electrons in the small active area of the chip) to provide lasing action. Some of the photons emitted by the spontaneous action are trapped in the Fabry-Perot cavity, reflecting back and forth from end mirror to end mirror. These photons have an energy level equal to the band gap of the laser materials. If one of these photons influences an excited electron, the electron immediately recombines and gives off a photon. Remember that the wavelength of a photon is a measure of its energy. Since the energy of the stimulated photon is equal to the original, stimulating photon, its wavelength is equal to that of the original, stimulating photon. The photon created is a *duplicate of the first photon: It has the same wavelength, phase, and direction of travel.* In other words, the incident photon has stimulated the emission of another photon. Amplification has occurred, and emitted photons have stimulated further emission.

The high drive current in the chip creates population inversion. *Population inversion* is the state in which a high percentage of the atoms moves from the ground state to the excited state so that a great number of free electrons and holes exists in the active area around the junction. When population inversion is present, a photon is more likely to stimulate emission than be absorbed. Only above the threshold current does population inversion exist at a level sufficient to allow lasing.

Although some of the photons remain trapped in the cavity, reflecting back and forth and stimulating further emissions, others escape through the two cleaved end faces in an intense beam of light. Since light is coupled into the fiber only from the ''front'' face, the ''rear'' face is often coated with a reflective material to reduce the amount of light emitted. Light from the rear face can also be used to monitor the output from the front face. Such monitoring can be used to adjust the drive current to maintain constant power level on the output.

Thus, the laser differs from an LED in that laser light has the following attributes:

- *Nearly monochromatic:* The light emitted has a narrow band of wavelengths. It is nearly *monochromatic*—that is, of a single wavelength. In contrast to the LED, laser light is not continuous across the band of its special width. Several distinct wavelengths are emitted on either side of the central wavelength.
- *Coherent:* The light wavelengths are in phase, rising and falling through the sine-wave cycle at the same time.
- *Highly directional:* The light is emitted in a highly directional pattern with little divergence. *Divergence* is the spreading of a light beam as it travels from its source.

SAFETY

Fiber-optic sources—lasers and high-radiance LEDs, especially—emit intense infrared light invisible to the human eye. Such radiation can permanently damage the retina and produce visual impairment or blindness. Never look directly into a source or into the end of a fiber energized by a source. Indeed, do not look into a source or fiber unless you are *positive* it is not energized. Remember, you cannot see whether a source or fiber is energized or not. Be careful.

SOURCE CHARACTERISTICS

This section describes some of the main characteristics of interest for sources. In doing so, it compares LEDs and lasers. These characteristics help determine the suitability of an LED or a laser for a given application. Table 8–1 provides a brief comparison of laser and LED characteristics.

OUTPUT POWER

Output power is the optical power emitted at a specified drive current. As shown in Figure 8–7, an LED emits more power than a laser operating below the threshold. Above the lasing threshold, the laser's power increases dramatically with increases with drive current. In general, the output power of the device is in the following decreasing order: laser, edge-emitting LED, surface-emitting LED.

OUTPUT PATTERN

The output pattern of the light is important in fiber optics. As light leaves the chip, it spreads out. Only a portion actually couples into the fiber. A smaller output pattern allows more light to be coupled into the fiber. A good source should have a small emission diameter and a small NA. The emission diameter defines how large the area of emitted light is. The NA defines at what angles the light

Characteristic	LED	Laser
Output power	Lower	Higher
Speed	Slower	Faster
Output pattern (NA)	Higher	Lower
Spectral width	Wider	Narrower
Single-mode compatibility	No	Yes
Ease of Use	Easier	Harder
Lifetime	Longer	Long
Cost	Lower	Higher

TABLE 8–1 Relative characteristics: LED and laser

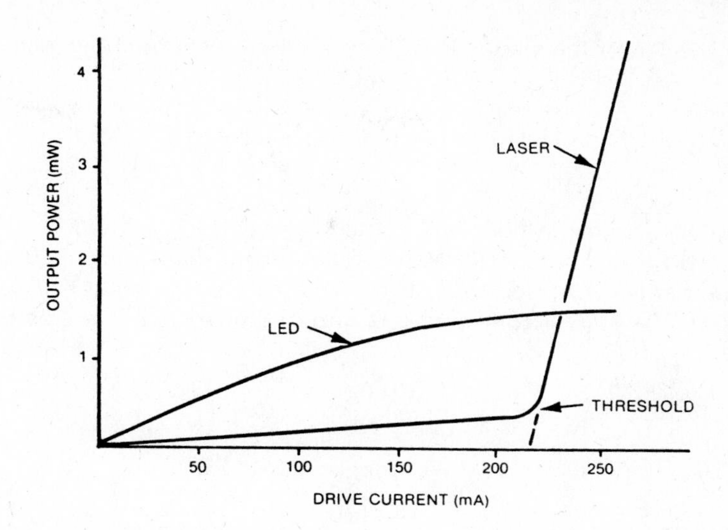

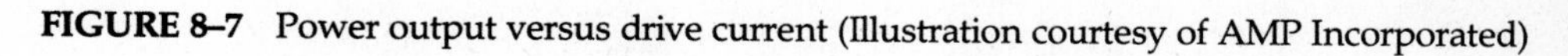
FIGURE 8–7 Power output versus drive current (Illustration courtesy of AMP Incorporated)

is spreading out. If either the emitting diameter or the NA of the source is larger than those of the receiving fiber, some of the optical power will be lost. Figure 8–8 shows typical emission patterns for an LED and a laser.

The loss of optical power from mismatches in NA and diameter between the source and the core of multimode fiber is as follows: When the diameter of the source is greater than the core diameter of the fiber, the mismatch loss is

$$\text{loss}_{\text{dia}} = 10 \log_{10} \left(\frac{\text{dia}_{\text{fiber}}}{\text{dia}_{\text{source}}} \right)^2$$

No loss occurs when the core diameter of the fiber is larger.

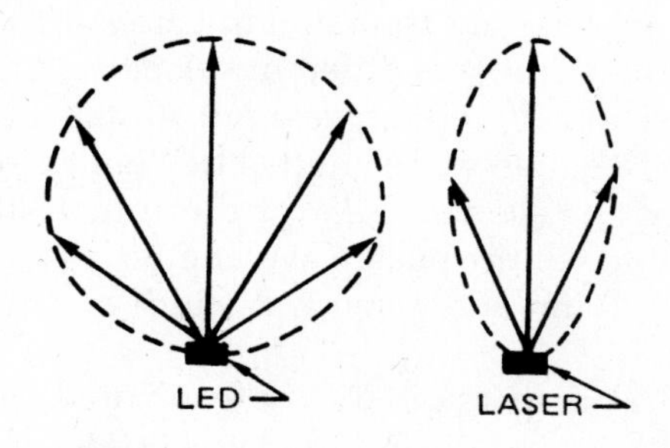

FIGURE 8–8 Emission patterns

When the NA of the source is larger than the NA of the fiber, the mismatch loss is

$$loss_{NA} = 10 \log_{10} \left(\frac{NA_{fiber}}{NA_{source}} \right)^2$$

No loss occurs when the fiber NA is the larger.

Consider, for example, a source with an output diameter of 100 μm and an NA of 0.30 that is connected to a fiber with a core diameter of 62.5 μm and an NA of 0.275. The losses from diameter and NA mismatches are as follows:

$$loss_{dia} = 10 \log_{10} \left(\frac{62.5}{100} \right)^2$$

$$= 10 \log_{10} (0.390625)$$

$$= -4.1 \text{ dB}$$

$$loss_{NA} = 10 \log_{10} \left(\frac{0.275}{0.30} \right)^2$$

$$= 10 \log_{10} (0.934444)$$

$$= -0.8 \text{ dB}$$

The total mismatch loss is 4.9 dB. If the source output power is 800 μW, only about 260 μW are coupled into the fiber.

A single-mode fiber requires a laser source. The laser provides a small, intense beam of light compatible with the small core of the fiber. The laser's output is elliptical, rather than circular.

SPECTRAL WIDTH

As the discussion of material dispersion in Chapter 6 emphasized, different wavelengths travel through a fiber at different velocities. The dispersion resulting from different velocities of different wavelengths limits bandwidth. Lasers and most LEDs do not emit a single wavelength; they emit a range of wavelengths. This range is known as the *spectral width* of the source. It is measured at 50% of the maximum amplitude of the peak wavelength. For example, if a source has a peak wavelength of 820 nm and a spectral width of 30 nm, its output ranges from 805 to 835 nm.

At 850 nm, material dispersion can be estimated as 0.1 ns/km for each nanometer of source spectral width. An LED with a 30-nm spectral width results in 3 ns of dispersion over a 1-km run.

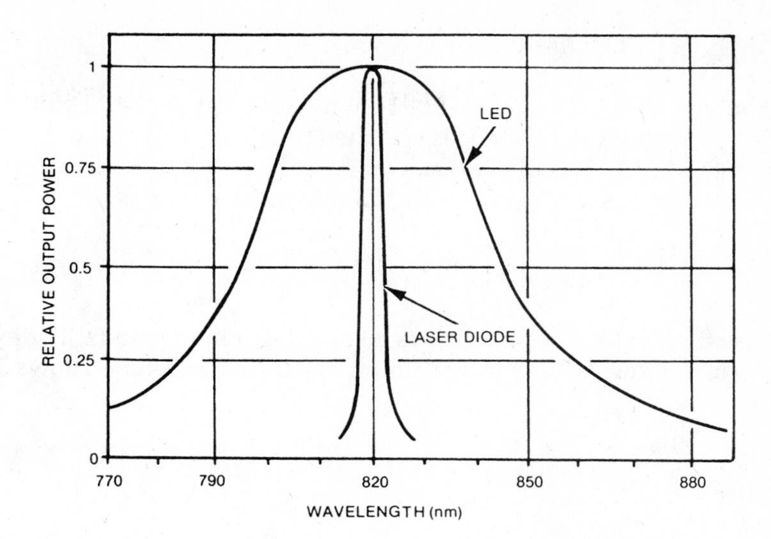

FIGURE 8–9 Typical spectral widths (Illustration courtesy of AMP Incorporated)

Figure 8–9 shows that the spectral width of a laser is substantially narrower than that of an LED. The spectral width of a laser is 2 to 5 nm, but that of an LED is tens of nanometers. As a rule of thumb, spectral width is not important in fiber-optic links running under 100 MHz for only a few kilometers. Spectral width is especially important in high-speed, long-distance, single-mode systems because the resulting dispersion is the principal limiting factor on system speed. Remember that the practical bandwidth of a single-mode fiber is specified as dispersion in picoseconds per kilometer per *nanometer of source width* (ns/km/nm).

Because laser spectral width is a main limiting characteristic of long-distance, high-speed, single-mode systems, great effort in recent years has been expended to devise and to build reliable single-wavelength laser diodes. Such devices use advanced structures to promote the center wavelength and to suppress others. Two examples of such devices are the *distributed-feedback laser* and the *cleaved-coupled cavity* (C^3) laser. The distributed-feedback laser uses an internal grating to limit the photons that can resonate (reflect from the two end walls of the cavity) to one wavelength. Thus, all stimulated emissions have the same wavelength. The C^3 laser uses a laser-diode chip separated into two sections by a small gap. Each section is operated separately, and the optical interaction between the two sections purifies the output by suppressing some wavelengths and promoting one wavelength.

SPEED

The source must turn on and off fast enough to meet the bandwidth requirements of the system. Source speed is specified by rise and fall times. Lasers have rise times of less than 1 ns, whereas slower LEDs have rise times of nanoseconds. A rough approximation of bandwidth for a given rise time is

$$BW = \frac{0.35}{t_r}$$

where a rise time in nanoseconds gives a bandwidth in gigahertz. For example, a 1-ns rise time allows 350-MHz operation, and a 5-ns rise time allows 70-MHz operation.

LIFETIME

The expected operating lifetime of a source runs into the millions of hours. Over time, however, the output power decreases because of increasing defects in the device's crystalline structure. The lifetime of the source is typically taken to end when the peak output power is reduced 50% or 3 dB. An LED emitting a peak power of 1 mW, for example, is considered at the end of its lifetime when its peak power becomes 500 μW.

EASE OF USE

Although a laser provides better optical performance than an LED, it is also more expensive, less reliable, and harder to use. Lasers have a shorter expected lifetime than LEDs. They also require more complex transmitter circuits. For example, the output power of a laser can change significantly with temperature. Maintaining proper output levels over the required temperature range requires circuitry that detects changes in output and adjusts the drive current accordingly. One method of maintaining these levels is with a photodiode to monitor the light output on the back facet of the laser. The current from the photodiode changes with variations in light output. These variations provide feedback to adjust the laser drive current.

PACKAGING

A central issue in packaging the source for use in fiber optics is to couple as much optical power into the fiber as possible. The optical characteristics of the source must be compatible with the optical characteristics of the fiber. This section looks at many of the techniques used to package the source for fiber-optic systems.

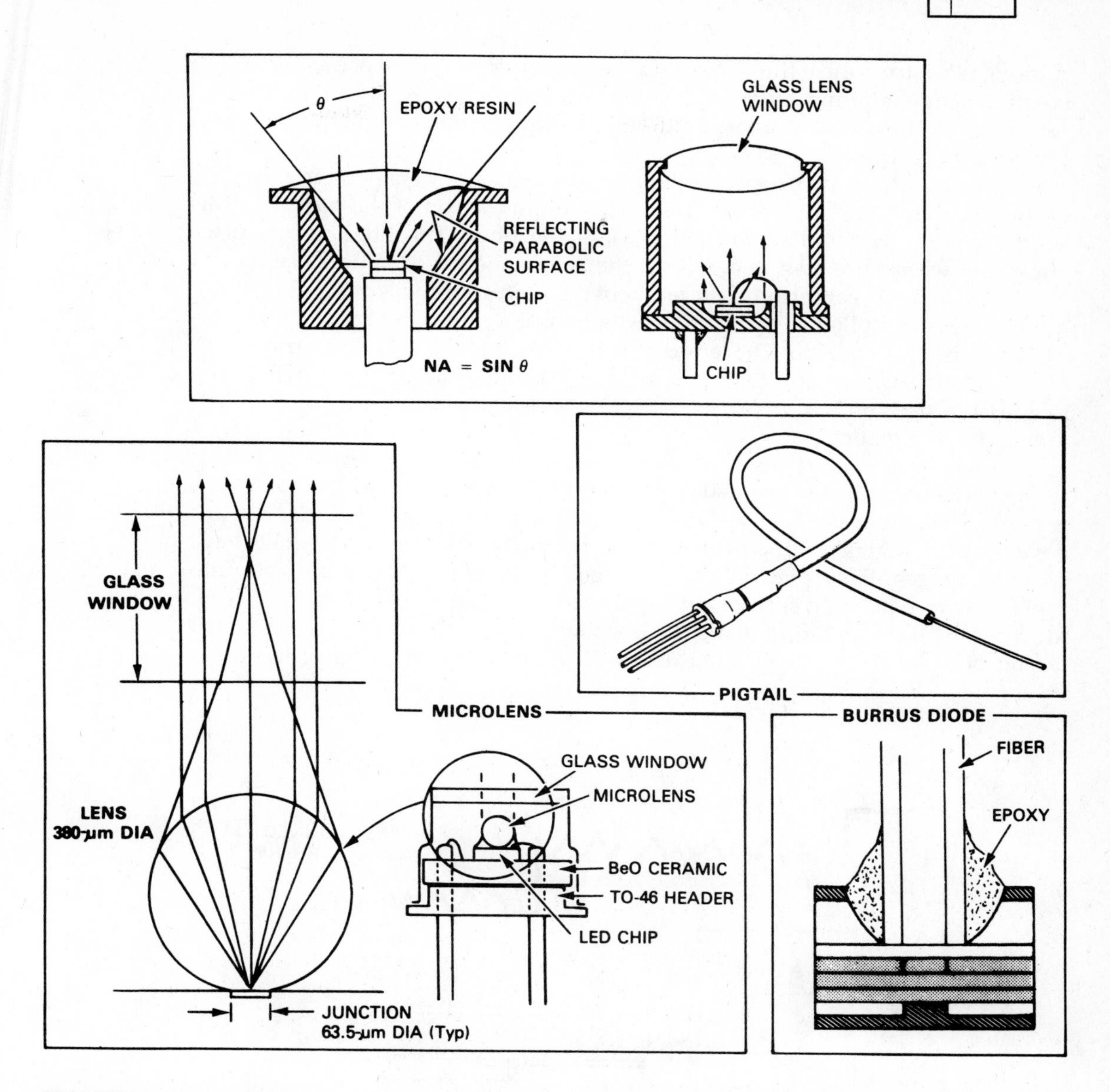

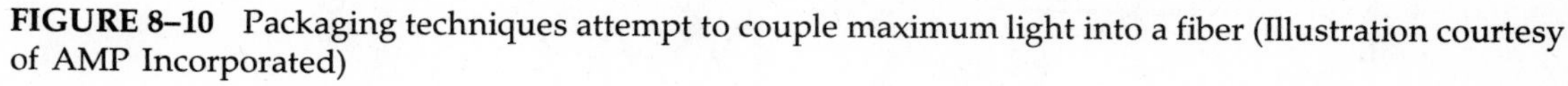

FIGURE 8–10 Packaging techniques attempt to couple maximum light into a fiber (Illustration courtesy of AMP Incorporated)

In a surface-emitting LED with a large emitting diameter, only a small portion of this optical energy is coupled into the fiber. In addition, the angles at which light emits tend to be very broad, so that the light spreads out quickly. This fact leads to an important point about sources: The total output power of a chip and the power that is usably coupled into the fiber can vary dramatically.

Although the structure of the semiconductor allows better control of output pattern (as in an edge-emitting diode and laser), source packaging also allows better coupling of light into the fiber. Figure 8–10 shows several techniques used to control the light output.

A *microlensed* device uses a small glass bead epoxied directly to the chip. The bead focuses the light in a nearly uniform spot on the top of the package. This spot is usually larger than the diameter of the fiber. The fiber can be place anywhere within the spot and receive the same amount of optical energy.

The lens bead can also be placed above the chip to focus the light. In the example shown in Figure 8–11, the lens is part of a mounting to which a fiber-optic connector is attached. The lens directs the light into the fiber in the connector.

Parabolic surfaces and lens caps help to collimate the light, reducing both its emission area and NA. In collimated light, all rays travel parallel to one another, rather than spreading.

Pigtails have a short length of optical fiber as part of the device. A Burrus LED has the fiber directly epoxied into a well etched into the chip. The advantage is to bring the fiber very close to the active region of the device. Other devices have the fiber pigtails aligned very close to the surface of the chip. Pigtails have two advantages. First, placing the fiber close to the chip couples the light into the fiber before the light spreads out. The farther away the fiber is from the chip, the more the light spreads out and the less the light that enters the fiber. Second, the output power of the pigtailed device is the power at the end of the fiber.

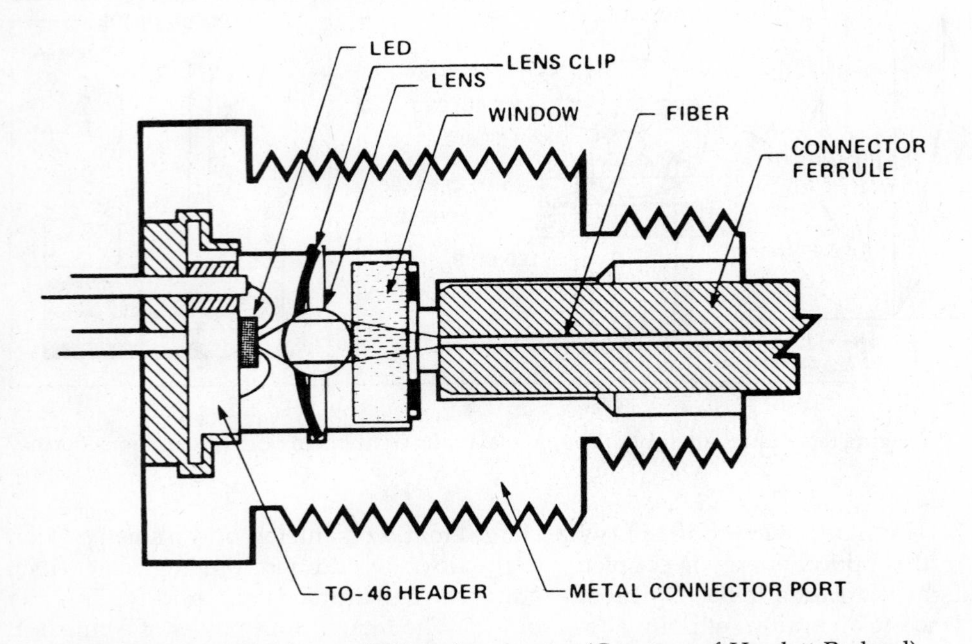

FIGURE 8–11 Microlenses LED and connector (Courtesy of Hewlett-Packard)

Second, the output power of the pigtailed device is the power at the end of the fiber. This characteristic simplifies system calculations because the emission pattern from the pigtail is more easily specified.

The significance of source packaging can be seen in the offering of one manufacturer. The same LED chip is offered in three different packages. The specified output for each package is as follows:

- Unpigtailed
  300 μW
- Pigtailed with 100/140 fiber
  250 μW
- Pigtailed with 50/125 fiber
  50 μW

The difference in power in the two pigtailed versions results from the greater difficulty of coupling power into the small core and NA of the 50/125 fiber.

Sources are often packaged in a receptacle designed to mate with a specific type of connector. Figure 8–12 shows a variety of such receptacles. Such packaged solutions offer the user several benefits. First, the manufacturer can specify the amount of fiber coupled into a particular fiber size, which eliminates the need to calculate loss values for the source-to-fiber connection. Second, it simplifies

FIGURE 8–12 Source receptacle packages (Courtesy of AMP Incorporated)

	LED			LED			LASER		
	Min	Typ	Max	Min	Typ	Max	Min	Typ	Max
Coupled Power (μW)									
50-μm fiber	30	55	80	25	35	—	—	—	—
62.5-μm fiber	44	100	175	50	75	—	—	—	—
Single-mode fiber	—	—	—	0.5	1	—	1	1000	—
Wavelength (nm)	820	835	850	1290	1320	1350	1280	1310	1330
FWHM Spectral Width (nm)	—	—	75	—	—	170	—	—	5
Risetime (ns)	1.5	3.5	4	2	2.5	4	—	0.3	—

TABLE 8–2 Typical LED and laser characteristics

design since the module is ready to mount on a pc board. Receptacles are available for popular connector types.

SOURCE EXAMPLES

Table 8–2 provides typical specifications for LEDs and lasers.

SUMMARY

- The source is the electro-optic transducer in which a forward current results in emission of light.
- An LED emits light spontaneously; a laser uses stimulated emission to achieve higher outputs above the threshold current.
- A laser has a narrower spectral width than an LED.
- A laser operates faster than an LED.
- Single-mode fibers require a laser.
- LEDs are less expensive and easier to operate than lasers.
- Source packaging is important to efficient coupling of light into a fiber.

REVIEW QUESTIONS

1. What is the purpose of the source?
2. Name the two main types of fiber-optic sources.
3. What is emitted when a free electron recombines with a hole in a semiconductor source?
4. From what phrase is the word *laser* derived?
5. List three packaging methods used to make coupling of light into a fiber more efficient.
6. If a source has a rise time of 7 ns, what is its approximate bandwidth?
7. List three characteristics of laser light that differentiate it from LED light.
8. What relationship between source NA and fiber NA results in NA mismatch loss?

CHAPTER 9

Detectors

The *detector* performs the opposite function from the source: It converts optical energy to electrical energy. The detector is an optoelectronic transducer. A variety of detector types is available. The most common is the *photodiode,* which produces current in response to incident light. Two types of photodiodes used extensively in fiber optics are the pin photodiode and the avalanche photodiode. This chapter describes photodiode detectors and their characteristics most important to our study of fiber optics.

PHOTODIODE BASICS

Chapter 8 discussed the theory of energy bands in semiconductor material. In moving from the conduction band to the valence band, by recombination of electron-hole pairs, an electron gives up energy. In an LED, this energy is an emitted photon of light with a wavelength determined by the band gap separating the two bands. Emission occurs when current from the external circuit passes through the LED.

With a photodiode, the opposite phenomenon occurs: light falling on the diode creates current in the external circuit. Absorbed photons excite electrons from the valence band to the conduction band, a process known as *intrinsic absorption.* The result is the creation of an electron-hole pair. These carriers, under the influence of the bias voltage applied to the diode, drift through the material and induce a current in the external circuit. For each electron-hole pair thus created, an electron is set flowing as current in the external circuit.

PN PHOTODIODE

The simplest photodiode is the pn photodiode shown in Figure 9–1. Although this type of detector is not widely used in fiber optics, it serves to illustrate the basic ideas of semiconductor photodetection. Other devices—the pin and avalanche photodiodes—are designed to overcome the limitations of the pn diode.

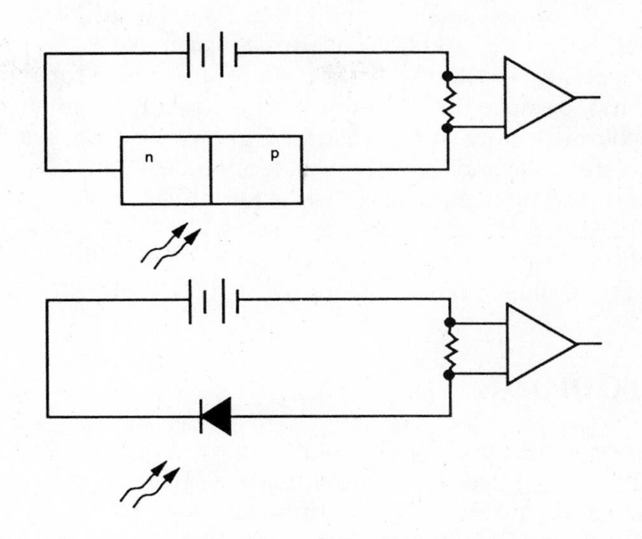

FIGURE 9–1 Pn photodiode

The pn photodiode is a simple pn device. When it is reverse biased (negative battery terminal connected to p-type material), very little current flows. The applied electric field creates a depletion region on either side of the pn junction. Carriers—free electrons and holes—leave the junction area: In other words, electrons migrate toward the negative terminal of the device (toward the positive terminal of the battery) and holes toward the positive terminal (negative terminal of the battery). Because the depletion region has no carriers, its resistance is very high, and most of the voltage drop occurs across the junction. As a result, electrical forces are high in this region and negligible elsewhere.

An incident photon absorbed by the diode gives a bound electron sufficient energy to move from the valence band to the conduction band, creating a free electron and a hole. If this creation of carriers occurs in the depletion region, the carriers quickly separate and drift rapidly toward their respective regions. This movement sets an electron flowing as current in the external circuit. When the carriers reach the edge of the depletion region, where electrical forces are small, their movement, and hence their external current, ceases.

When electron-hole creation occurs outside of the depletion region, the carriers move slowly toward the depletion region. Many carriers recombine before reaching it. External currents are negligible. Those carriers remaining and reaching the depleted area are swiftly swept across the junction by the large electrical forces in the region to produce an external electrical current. This current, however, is delayed with respect to the absorption of the photon that created the carriers because of the initial slow movement of carriers toward the depletion region.

Current, then, will continue to flow after the light is removed. This slow response, due to slow diffusion of carriers, is called *slow tail response.*

Two characteristics of the pn diode make it unsuitable for most fiber-optic applications. First, because the depletion area is a relatively small portion of the diode's total volume, many of the absorbed photons do not result in external current: The created hole and free electrons recombine before they cause external current. The received optical power must be fairly high to generate appreciable current. Second, the slow tail response from slow diffusion makes the diode too slow for medium- and high-speed applications. This slow response limits operations to the kilohertz range.

PIN PHOTODIODE

The structure of the *pin diode* is designed to overcome the deficiencies of its pn counterpart. The depletion region is made as large as possible by the pin structure shown in Figure 9–2. A lightly doped intrinsic layer separates the more heavily doped p-type and n-type materials. *Intrinsic* means that the material is not doped to produce n-type material with free electrons or p-type material with holes. Although the intrinsic layer is actually lightly doped positive, the doping is light enough to allow the layer to be considered intrinsic—that is, neither strongly n-type nor p-type. The name of the diode comes from this layering of materials: *p*ositive, *i*ntrinsic, *n*egative—pin.

Since the intrinsic layer has no free carriers, its resistance is high, and electrical forces are strong within it. The resulting depletion region is very large in comparison to the size of the diode. The pin diode works like the pn diode. The large intrinsic layer, however, means that most of the photons are absorbed within the depletion region for better efficiency. The result is improved efficiency in incident photons, creating external current and faster speed. Carriers created within the depletion region are immediately swept by the electric field toward their p or n terminals.

A tradeoff exists in arriving at the best pin photodiode structure. Efficiency of photon-to-carrier conversion requires a thick intrinsic layer to increase the probability of incident photons creating electron-hole pairs in the depletion region. Speed, however, requires a thinner layer to reduce the transit time of the carriers swept across the region. Diode design involves balancing these opposing requirements to achieve the best balance between efficiency and speed.

AVALANCHE PHOTODIODE (APD)

For a pin photodiode, each absorbed photon ideally creates one electron-hole pair, which, in turn, sets one electron flowing in the external circuit. In this sense,

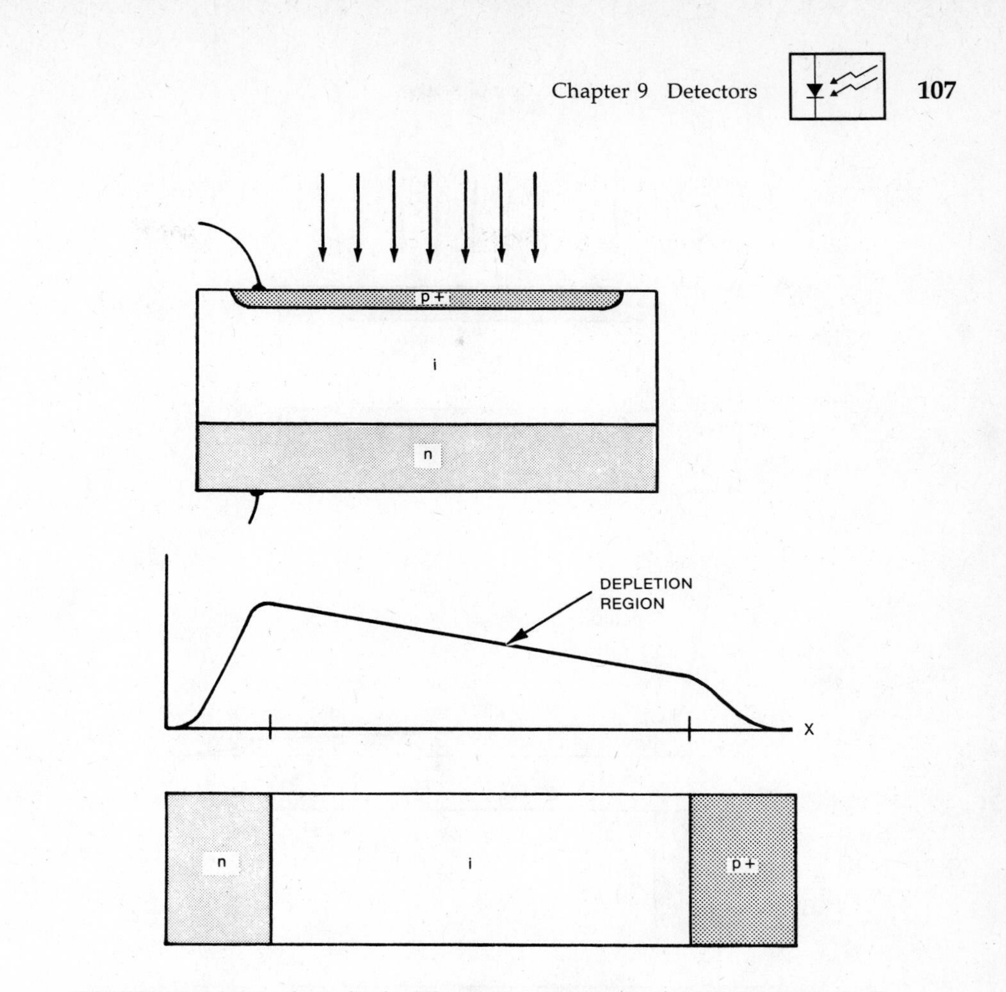

FIGURE 9–2 Pin photodiode (Illustration courtesy of AMP Incorporated)

we can loosely compare it to an LED. There is basically a one-to-one relationship between photons and carriers and current. Extending this comparison allows us to say that an avalanche photodiode resembles a laser, where the relationship is not one-to-one. In a laser, a few primary carriers result in many emitted photons. In an avalanche photodiode (APD), a few incident photons result in many carriers and appreciable external current.

The structure of the APD, shown in Figure 9–3, creates a very strong electrical field in a portion of the depletion region. *Primary carriers*—the free electrons and holes created by absorbed photons—within this field are accelerated by the field, thereby gaining several electron volts of kinetic energy. A collision of these fast carriers with neutral atoms causes the accelerated carrier to use some of its energy to raise a bound electron from the valence band to the conduction band. A free electron and hole appear. Carriers created in this way, through collision with a primary carrier, are called *secondary carriers.*

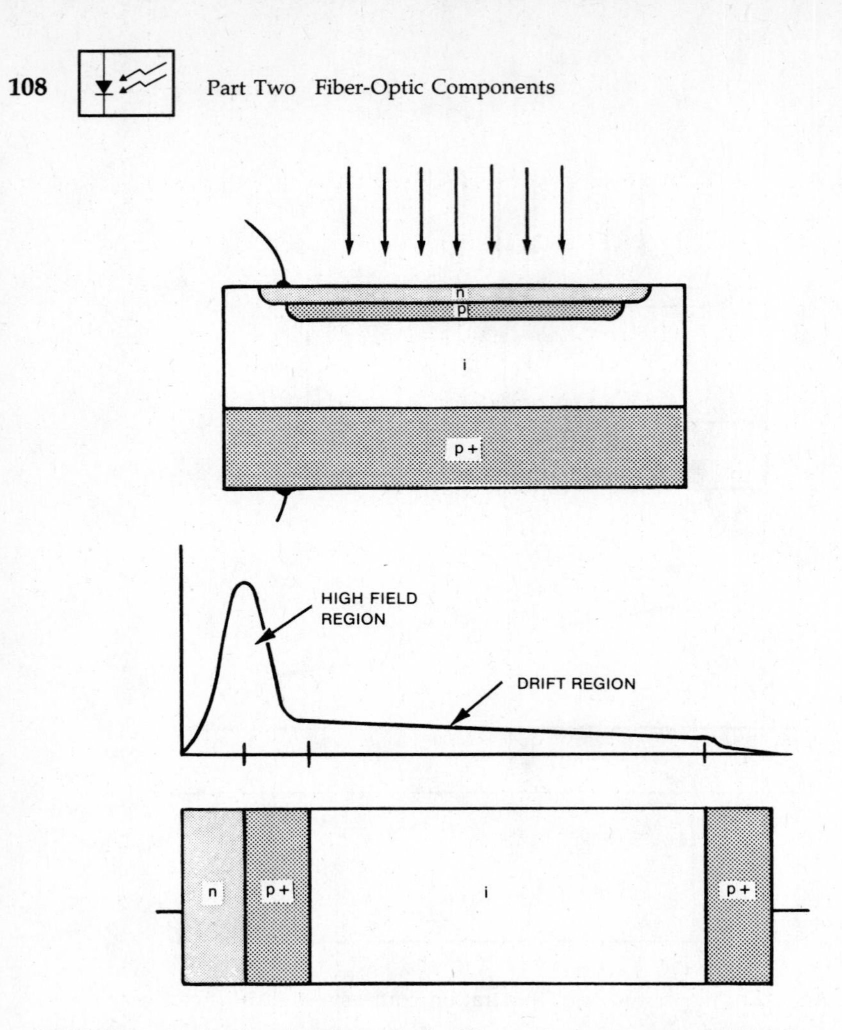

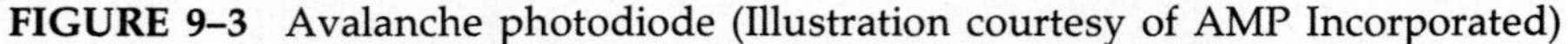

FIGURE 9–3 Avalanche photodiode (Illustration courtesy of AMP Incorporated)

This process of creating secondary carriers is known as *collision ionization.* A primary carrier can create several new secondary carriers, and secondary carriers themselves can accelerate and create new carriers. The whole process is called *photomultiplication,* which is a form of gain.

The number of electrons set flowing in the external circuit by each absorbed photon depends on the APD's multiplication factor. Typical multiplication ranges in the tens and hundreds. A multiplication factor of 70 means that, on the average, 70 external electrons flow for each photon. The phrase "on the average" is important. The multiplication factor is an average, a statistical mean. Each primary carrier created by a photon may create more or less secondary carriers and therefore external current.

For an APD with a multiplication factor of 70, for example, any given primary carrier may actually create 67 secondary carriers or 76 secondary carriers. This variation is one source of noise that limits the sensitivity of a receiver using an APD. We will discuss noise in fuller detail shortly.

The multiplication factor varies with the bias voltage. Because the accelerating forces must be strong enough to impart energies to the carriers, high bias voltages (several hundred volts in many cases) are required to create the high-field region. At lower voltages, the APD operates like a pin diode and exhibits no internal gain.

The avalanche breakdown voltage of an APD is the voltage at which collision ionization begins. An APD biased above the breakdown point will produce current in the absence of optical power. The voltage itself is sufficient to create carriers and cause collision ionization.

The APD is often biased just below the breakdown point, so any optical power will create a fast response and strong output. The tradeoffs are that dark current (the current resulting from generation of electron-hole pairs even in the absence of absorbed photons) increases with bias voltage, and a high-voltage power supply is needed.

NOISE

The purpose of the detector is to create an electrical current in response to incident photons. It must accept highly attenuated optical energy and produce a current. This current is usually feeble because of the low levels of optical power involved, often only tens or hundreds of nanowatts. Subsequent stages of the receiver amplify and possibly reshape the signal from the detector.

Noise is an ever-present phenomenon that seriously limits the detector's operation. *Noise* is any electrical or optical energy apart from the signal itself. The signal is wanted energy; noise is anything else—that is, unwanted energy. Although noise can and does occur in every part of a communication system, it is of greatest concern in the receiver input. The reason is that the receiver works with very weak signals that have been attenuated during transmission. Although very small compared to the signal levels in most circuits, the noise level is significant in relation to the weak detected signals. The same noise level in a transmitter is usually insignificant because signal levels are very strong in comparison. Indeed, the very limit of the diode's sensitivity is the noise. An optical signal that is too weak cannot be distinguished from the noise. To detect such a signal, we must either reduce the noise level or increase the power level of the signal.

The amplification stages of the receiver amplify both the signal and the noise. Although it is possible to use electronic circuits to filter out some types of noise, it is better to have the signal much stronger than the noise by either having a strong signal level or a low noise level. Having both is even better.

Several types of noise are associated with the photodetector itself and with the receiver. We have already mentioned multiplication noise in an APD, which arises because multiplication varies around a statistical mean. Two other types of noise most important to our understanding of photodiodes and fiber optics are shot noise and thermal noise.

SHOT NOISE

Shot noise arises from the discrete nature of electrons. Current is not a continuous, homogeneous flow. It is the flow of individual, discrete electrons. Keep in mind that a photodiode works because an absorbed photon creates an electron-hole pair that sets an external electron flowing as current. It is a three-step sequence: photon, electron-hole carriers, electron. The arrival and absorption of each photon and the creation of carriers are part of a random process. It is not a perfect homogeneous stream but a series of discrete occurrences. Therefore, the actual current fluctuates, as more or less electron-hole pairs are created in any given moment.

Shot noise exists even when no light falls on the detector. Even without light, a small trickle of current is generated thermally, increasing about 10% for every increase of 1°C. A typical dark current is 25 nA at 25°.

Shot noise is equal to

$$i_{sn}^2 = 2qiB$$

where q is the charge of an electron (1.6×10^{-19} coulomb), i is average current (including dark current and signal current), and B is the receiver bandwidth. The equation shows that shot noise increases with current and with bandwidth. Shot noise is at its minimum when only dark current exists (when i = dark current), and it increases with the current resulting from optical input. A detector with a dark current of 2 nA and operating at a bandwidth of 10 MHz has a shot noise of 80 pA:

$$\begin{aligned} i_{sn}^2 &= 2qiB \\ &= (2)(1.6 \times 10^{-19})(2 \times 10^{-9})(10 \times 10^{6}) \\ &= 6.4 \times 10^{-21} \\ i_{sn} &= 8 \times 10^{-11} \\ &= 80 \text{ pA} \end{aligned}$$

THERMAL NOISE

Thermal noise, also called *Johnson noise* or *Nyquist noise,* arises from fluctuations in the load resistance of the detector. The electrons in the resistor are not stationary: Their thermal energy allows them to constantly and randomly move around. At any given instant, the net movement can be toward one electrode or the other, so that randomly varying current exists. This random current will add to and distort the signal current from the photodiode.

Thermal noise is equal to

$$i_{tn}^2 = \frac{4kTB}{R_L}$$

where k is Boltzmann's constant (1.38×10^{-23} J/K), T is absolute temperature (kelvin scale), B is the receiver's bandwidth, and R_L is the load resistance.

Consider a 510-Ω load resistance operating at an absolute temperature of 298 K. Assume a bandwidth of 10 MHz. The thermal noise is

$$i_{tn}^2 = \frac{4kTB}{R_L}$$

$$= \frac{(4)(1.38 \times 10^{-23})(298)(10 \times 10^6)}{510}$$

$$= 3.23 \times 10^{-16}$$

$$i_{tn} = 1.79 \times 10^{-8}$$

$$= 18 \text{ nA}$$

Thermal and shot noise exist in the receiver independently of the arriving optical power. They result from the very structure of matter. They can be minimized by careful design of devices and circuits, but they cannot be eliminated. Any signal—optical, electrical, or human—must exist in the presence of noise. In the receiver stages following the detector, the signal and noise will be amplified. Therefore, the signal must be appreciably larger than the noise. If the signal power is equal to the noise power, the signal will not even be adequately detected. As a general rule, the optical signal should be twice the noise current to be adequately detected.

SIGNAL-TO-NOISE RATIO

Signal-to-noise ratio (SNR) is a common way of expressing the quality of signals in a system. SNR is simply the ratio of the average signal power to the average noise power from all noise sources.

$$\text{SNR} = \frac{S}{N}$$

In decibels, SNR equals

$$\text{SNR} = 10 \log_{10}\left(\frac{S}{N}\right)$$

If the signal current is 50 μW and the noise power is 50 nW, the ratio is 1000, or 30 dB.

A large SNR means that the signal is much larger than the noise. The signal power depends on the power of the arriving optical power. Different applications require different SNRs—that is, different levels of ''fidelity'' or freedom from distortion. The SNR required for a telephone voice channel is less than that required for a television signal, since a fair amount of noise on a telephone line will go unnoticed. We will also accept greater distortion in voices than we will accept in television picture quality. Further, a broadcast-quality television signal, which is the signal produced by the television networks, has a higher SNR than the television signal received in our homes. Why? The broadcast signal itself encounters noise during its transmission. It must, therefore, have a higher SNR so that the signal, after picking up noise during its transmission and reception, still has an SNR high enough to produce a clear, sharp picture.

BIT-ERROR RATE

For digital systems, bit-error rate (BER) usually replaces SNR as a measure of system quality. BER is the ratio of incorrectly transmitted bits to correctly transmitted bits. A ratio of 10^{-9} means that one wrong bit is received for every 1 billion bits transmitted. As with SNR, the required BER varies with the application. Digitally encoded telephone voices have a lower BER requirement than digital computer data, say 10^{-6} versus 10^{-9}. A few faulty bits will not cause noticeable distortion of a voice. A few faulty bits in computer data can cause significant changes in financial data or student grades. They can mean the difference between a program running successfully or crashing.

BER and SNR are related. A better SNR brings a better BER. BER, though, also depends on data-encoding formats and receiver designs. Techniques exist to detect and correct bit errors. We cannot easily calculate the BER from the SNR, because the relationship depends on several factors, including circuit design and bit-error correction schemes. In a given system, for example, an SNR of 22 dB may be required to maintain a BER of 10^{-9}, whereas an SNR of 17 dB brings a BER of 10^{-6}. In another design, however, the 10^{-9} BER might be achieved with an SNR of 18 dB.

DETECTOR CHARACTERISTICS

The detector characteristics of interest here are those that relate most directly to use in a fiber-optic system. These include the device's response to incident optical power and its speed. Since pin diodes are the most commonly used, we concentrate on those.

Responsivity

Responsivity (R) is the ratio of the diode's output current to input optical power and is given in amperes/watt (A/W). Optical power produces current. A pin photodiode typically has a responsivity of around 0.4 to 0.6 A/W. A responsivity of 0.6 A/W means that incident light having 50 μW of power results in 30 μA of current:

$$I_d = 50\ \mu W \times 0.6\ A/W = 30\ \mu A$$

where I_d is the diode current.

For an APD, a typical responsivity is 75 A/W. The same 50 μW of optical power now produces 3.75 mA of current.

$$I_d = 50\ \mu W \times 75\ A/W = 3750\ \mu A = 3.75\ mA$$

Responsivity varies with wavelength, so it is specified either at the wavelength of maximum responsivity or at a wavelength of interest, such as 850 nm or 1300 nm. Silicon is the most common material used for detectors in the 800- to 900-nm range. Its peak responsivity is 0.7 A/W at 900 nm. At the 850-nm wavelength, responsivity is still near the peak. In a fiber-optic system using plastic fibers, operation is usually at 650 nm in the visible spectrum. Here, the photodiode's responsivity drops to around 0.3 to 0.4 A/W.

Silicon photodiodes are not suitable for the longer wavelengths of 1300 nm and 1550 nm. Materials for long wavelengths are principally geranmium (Ge) and indium gallium arsenide (InGaAs). An InGaAs pin photodiode has a fairly broad and flat response curve. It does not peak nearly as sharply as does silicon's. Its responsivity of greater than 0.5 A/W from 900 to 1650 nm allows it to be used at both 1300 nm and 1550 nm. Figure 9–4 depicts typical responsivity curves for photodiodes.

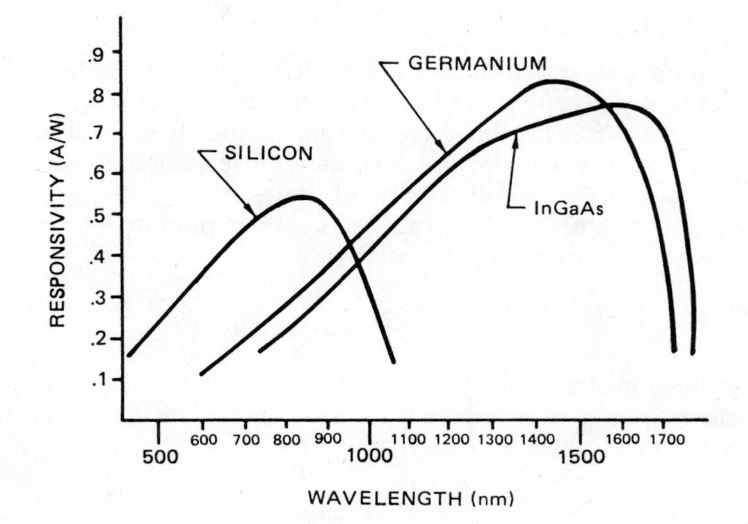

FIGURE 9–4 Responsivity

Quantum Efficiency (η)

Quantum efficiency is the ratio of primary electron-hole pairs (created by incident photons) to the photons incident on the diode material. It is expressed either as a dimensionless number or as a percentage. A quantum efficiency of 1, or 100%, means that every absorbed photon creates an electron-hole pair. A typical quantum efficiency of 70%, which means only 7 out of every 10 photons create carriers (electron current). Quantum efficiency applies to primary electrons created by photon absorption, not to secondary carriers created by collision ionization.

Quantum efficiency deals with the fundamental efficiency of the diode for converting photons into free electrons. Responsivity can be calculated from quantum efficiency:

$$R = \frac{\eta e \lambda}{hc}$$

where e is the charge of an electron, h is Planck's constant, and c is the velocity of light. Since e, c, and h are constants, responsivity is simply a function of quantum efficiency and wavelength.

Dark Current

As mentioned earlier, *dark current* is the thermally generated current in a diode. It is the lowest level of thermal noise. Dark current increases about 10% for every increase of 1°C. It is much lower in Si photodiodes used at shorter wavelengths than in Ge or InGaAs photodiodes used at longer wavelengths.

Minimum Detectable Power

The minimum power detectable by the detector determines the lowest level of incident optical power that the detector can handle. In simplest terms, it is related to the dark current in the diode, since the dark current will set the lower limit. Other noise sources are factors, including those associated with the diode and those associated with the first stage of the receiver.

The *noise floor* of a pin diode, which tell us the minimum detectable power, is the ratio of noise current to responsivity:

$$\text{Noise floor} = \frac{\text{noise}}{\text{responsivity}}$$

For initial evaluation of a diode, we can use the dark current to estimate the noise floor. Consider a pin diode with $R = 0.5\ \mu\text{A}/\mu\text{W}$ and a dark current of 2 nA. The minimum detectable power is

$$\text{Noise floor} = \frac{2 \times 10^{-12}\ \text{A}}{0.5\ \mu\text{A}/\mu\text{W}}$$

$$= 4 \times 10^{-12}\ \text{W}$$
$$= 4\ \text{nW}$$

More precise estimates must include other noise sources, such as thermal and shot noise. As discussed, the noise depends on current, temperature, load resistance, and bandwidth.

Response Time

Response time is the time required for the photodiode to respond to optical inputs and produce external current. As with a source, response time is usually specified as a rise time and a fall time, measured between the 10% and 90% points of amplitude. Rise times range from 0.5 ns to tens of nanoseconds. Rise time is limited to the transit speed of the carriers as they sweep across the depletion region. It is influenced by the bias voltage: higher voltages bring faster rise times. A pin diode, for example, might have a rise time of 5 ns at 15 V and 1 ns at 90 V.

The response time of the diode relates to its usable bandwidth. Bandwidth can be approximated from rise time:

$$\text{BW} = \frac{0.35}{t_r}$$

The bandwidth, or operating range, of a photodiode can be limited by either its rise time or its *RC* time constant, whichever results in the slower speed or bandwidth. The bandwidth of a circuit limited by the *RC* time constant is

$$\text{BW} = \frac{1}{2\pi R_L C_d}$$

where R_L is the load resistance and C_d is the diode capacitance. The rise time of the circuit is

$$t_r = 2.19\ R_L C_d$$

Figure 9–5 shows the equivalent circuit model of a pin diode. It consists of a current source in parallel with a resistance and a capacitance. It appears as a

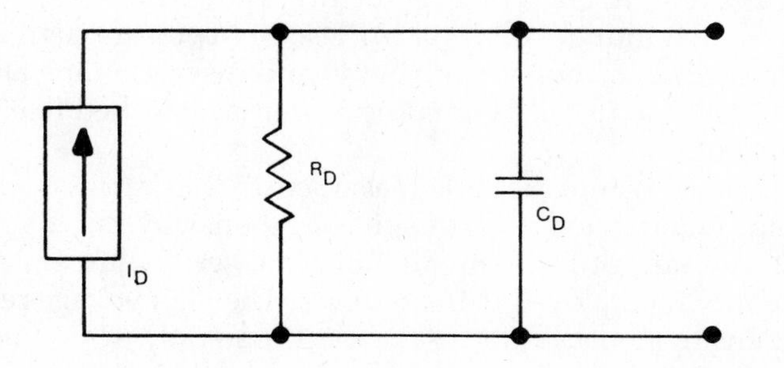

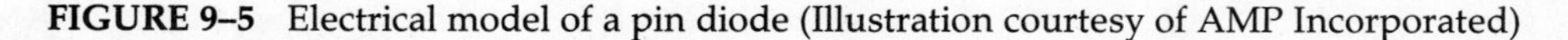
FIGURE 9–5 Electrical model of a pin diode (Illustration courtesy of AMP Incorporated)

low-pass filter, a resistor-capacitor network that passes low frequencies and attenuates high frequencies. The *cutoff* frequency, which is the frequency that is attenuated 3 dB or 50%, marks the bandwidth. Frequencies higher than cutoff are eliminated.

Diodes for high-speed operation must have capacitances of only a few picofarads or less. The capacitance in a pin diode is mainly the junction capacitance formed at the junctions of p, i, and n layers, as well as any capacitance contributed by the packaging structure and mounting.

Consider a photodiode with a rise time of 1 ns and a capacitance of 2 pF. Its approximate bandwidth, based on rise time, is

$$\begin{aligned} BW &= \frac{0.35}{1 \text{ ns}} \\ &= 0.35 \text{ GHz} = 350 \text{ MHz} \end{aligned}$$

To ensure that the *RC* time constant does not lower the bandwidth, we must determine the highest resistor value usable:

$$\begin{aligned} BW &= \frac{1}{2\pi R_L C_d} \\ 350 \times 10^6 \text{ Hz} &= \frac{1}{(2)(3.1415927)(2 \times 10^{-12})(R_L)} \\ R_L &= 227\ \Omega \end{aligned}$$

Thus, a standard 220-Ω resistor works, although in practice a resistor with one quarter of this value is often chosen.

Bias Voltage

Photodiodes require bias voltages ranging from as low as 5 V for some pin diodes to several hundred volts for APDs. Bias voltage significantly affects operation, since dark current, responsivity, and response time all increase with bias voltage. APDs are usually biased near their avalanche breakdown point to ensure fast response.

As shown in Figure 9–6, the APD is also sensitive to variations in temperature. The bias voltage required to maintain a given responsivity varies significantly with temperature. The output from an APD becomes erratic unless extensive compensating circuitry is employed in the receiver. The high voltage requirement and temperature sensitivity mean that APDs are chosen only when their responsivity and gain justify their complex application requirements.

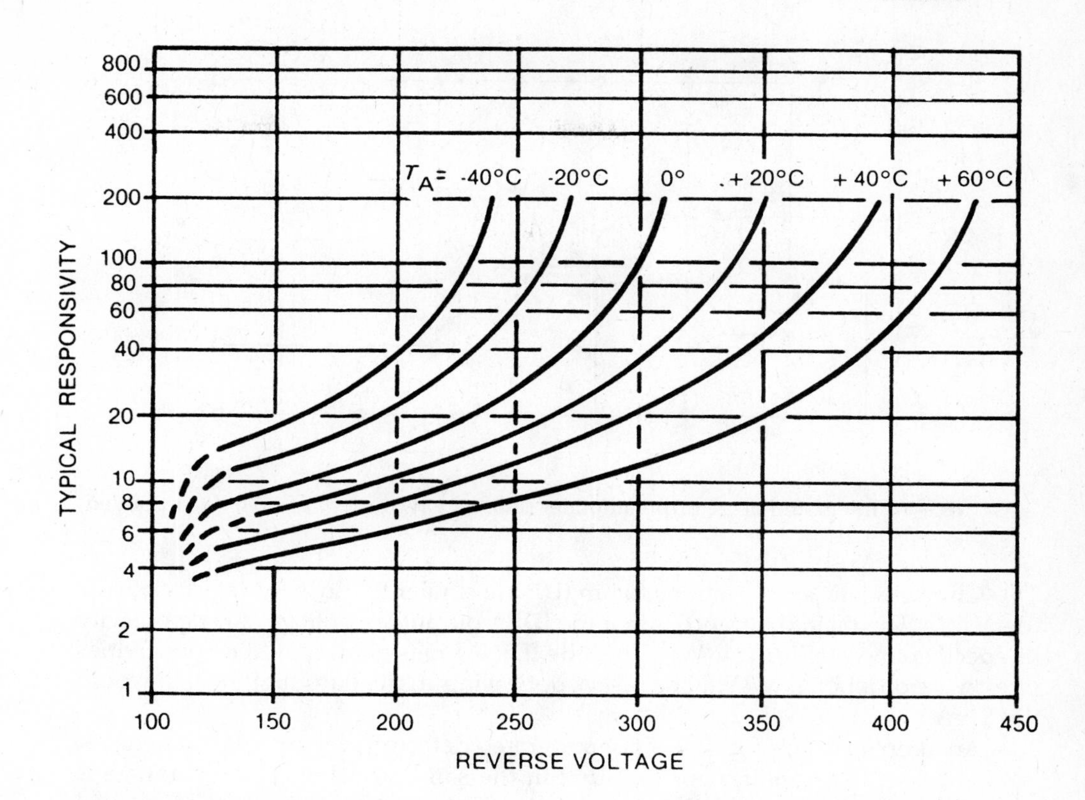

FIGURE 9–6 Effects of bias voltage and temperature on responsivity of an APD (Illustration courtesy of AMP Incorporated)

INTEGRATED DETECTOR/PREAMPLIFIER

The *integrated detector/preamplifier* (IDP) is an alternative to pin photodiodes. Noise that limits receiver operation can occur between the diode itself and the first receiver stage. The electrical leads of the diode may be sensitive to surrounding EMI and act as an antenna, picking up noise that will be coupled along with the noise into the amplifier.

To reduce these noise sources, the IDP has a *transimpedance amplifier* incorporated on the same semiconductor chip as the photodetector. The transimpedance amplifier performs both amplification and current-to-voltage conversion. The IDP, then, is an integrated circuit having both a photodiode and a transimpedance amplifier. (The amplifier can also be placed in the detector package as a separate chip.) Figure 9–7 shows a schematic of an integrated IDP.

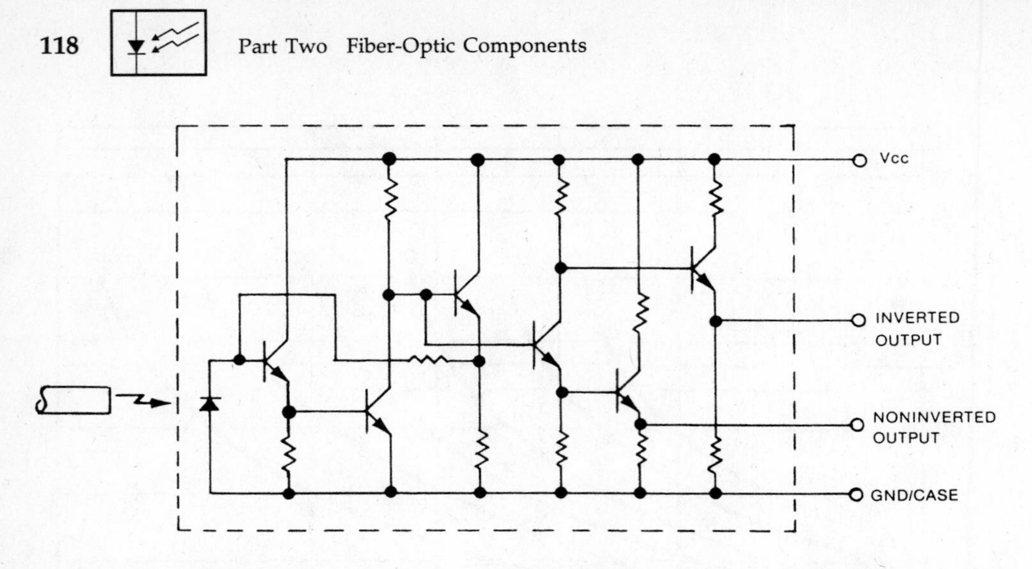

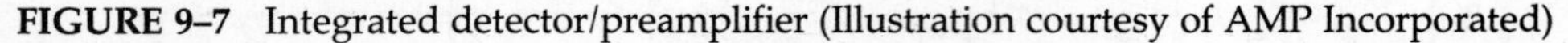
FIGURE 9–7 Integrated detector/preamplifier (Illustration courtesy of AMP Incorporated)

Characteristic specifications for an IDP are similar to those for regular photodetectors. The main difference is that the IDP's output is a voltage, so responsivity is specified in volts/watt (V/W). Since the IDP exhibits gain, typical responsivities are on the order of 40 V/W. For a 50-μW optical input, the output of the IDP would be 2 mV.

An alternative approach is a nonintegrated detector/preamp. Here a detector and separate preamplifier are included in the same package. The advantage is the same as for the integrated device: to have the two units close together to avoid noise problems.

PACKAGING

The packaging of the detector is similar to that of the source: TO series cans, pigtailed devices, and microlensed devices predominate. See the discussion of source packaging in Chapter 8 for further details.

As with sources, two main sources of loss in coupling light from a fiber into the detector results from mismatches in diameter and NA: When $\text{dia}_{\text{det}} < \text{dia}_{\text{fiber}}$,

$$\text{loss}_{\text{dia}} = 10 \log_{10} \left(\frac{\text{dia}_{\text{det}}}{\text{dia}_{\text{fiber}}} \right)^2$$

When $\text{NA}_{\text{det}} < \text{NA}_{\text{fiber}}$,

$$\text{loss}_{\text{NA}} = 10 \log_{10} \left(\frac{\text{NA}_{\text{det}}}{\text{NA}_{\text{fiber}}} \right)^2$$

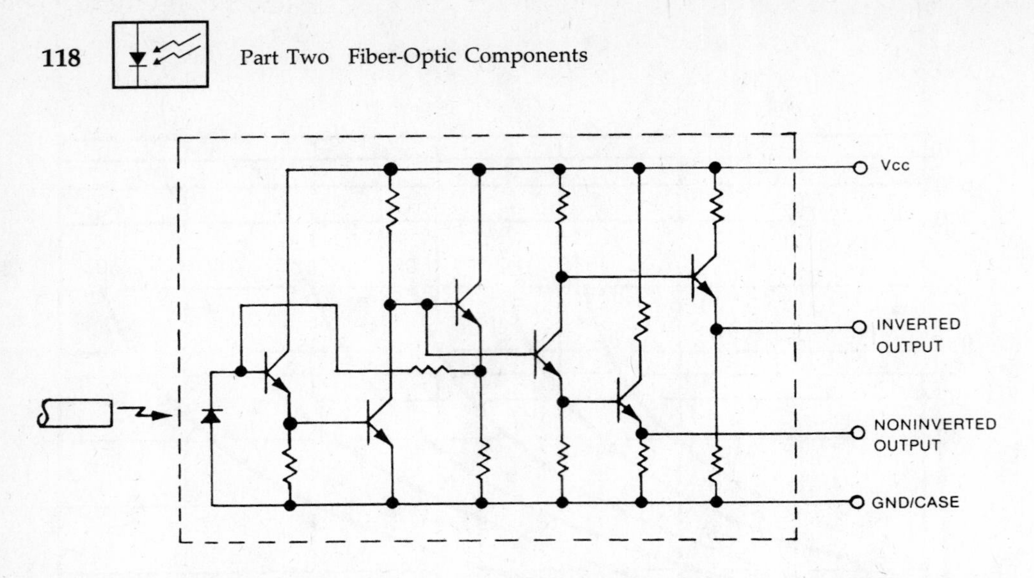

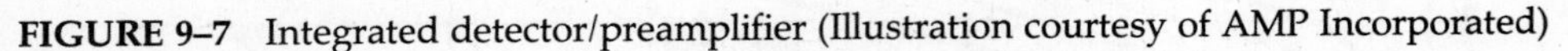

	Pin	Pin/Preamp	APD
Responsivity	.80 μA/μW	2 mV/μW	70 μA/μW
Spectral Response (nm)	1150–1600	1150–1600	1150–1600
Dark Current (nA)	2		5
Capacitance (pF)	1.5		4
Risetime (ns)	0.5 max		0.5

TABLE 9–1 Typical detector characteristics

Since detectors can be easily manufactured with large active diameters and wide angles of view, such mismatches are less common than with sources. Other losses occur from Fresnel reflections and mechanical misalignment between the connector and diode package.

Detectors are packaged in the same receptacles as sources (see Figure 8–12).

DETECTOR SPECIFICATIONS

Table 9–1 provides typical specifications for typical pin photodiodes and APDs.

SUMMARY

- The detector is the optoelectronic transducer that converts optical power into current.
- The two most common detectors used in fiber-optic communications are the pin photodiode and avalanche photodiode.
- Responsivity expresses the ratio of output current to optical power.
- An APD provides internal gain, so its responsivity is much higher than that of a pin diode. The APD is also more difficult to apply.
- Noise sets the lower limit of detectable optical power.
- Two types of noise associated with a detector are shot noise and thermal noise.
- SNR and BER are two methods of expressing the ''quality'' of a signal in a system.
- The response speed of a detector can be limited by its rise time or by its *RC* time constant.
- An IDP is a detector package containing both a pin photodiode and a transimpedance amplifier.

? REVIEW QUESTIONS

1. What is the purpose of the detector?
2. Name the two types of fiber-optic sources.
3. Name the three layers of a pin photodiode. What is the purpose of the middle layer?
4. Name the type of noise that results from current flowing as discrete electrons.
5. Name the type of noise that results from thermal energy in the load resistor.
6. Calculate the current that results from a received optical power of 750 nW and a responsivity of 0.7 A/W.
7. Name two factors that limit the response speed of a detector.
8. Calculate the rise time of a detector circuit having a capacitance of 3 pF and a load resistance of 160 Ω.
9. What distinguishes an IDP from a pin photodiode?
10. Give two reasons why a pin photodiode is used more often than an APD.

Transmitters and Receivers

In Chapter 1, we saw that a fiber-optic link contains a transmitter, optical cable, and receiver. We have looked at fiber-optic cables, sources, and detectors, the electro-optic transducers that provide the bridge between the optical and electronic parts of a fiber-optic system.

BASIC TRANSMITTER CONCEPTS

Figure 10–1 shows a basic block diagram of a transmitter. It contains a driver and a source. The input to the driver is the signal from the equipment being served. The output from the driver is the current required to operate the source.

Most electronic systems operate on standard, well-defined signal levels. Television video signals use a 1-V peak-to-peak level. Digital systems use different standards, depending on the type of logic circuits used in the system. These logic circuits define the levels for the highs and lows that represent the 1s and 0s of digital data.

Currently, the most common digital logic is transistor-transistor logic (TTL). TTL uses 0.5 V for a low signal and 5 V for a high signal. Although TTL is a fast logic, allowing it to be used in many applications, an even faster logic is emitter-coupled

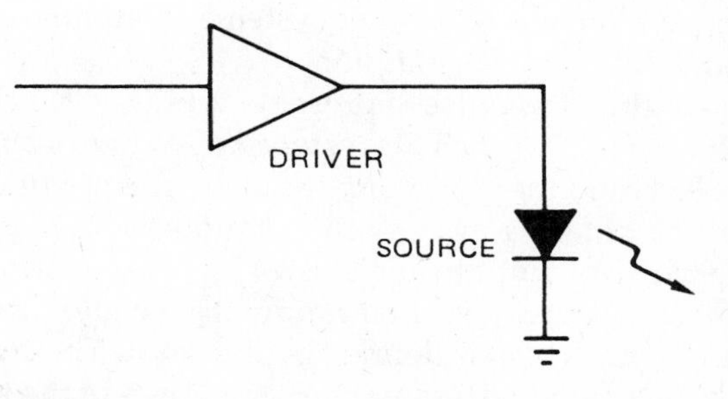

FIGURE 10–1 Basic transmitter block diagram

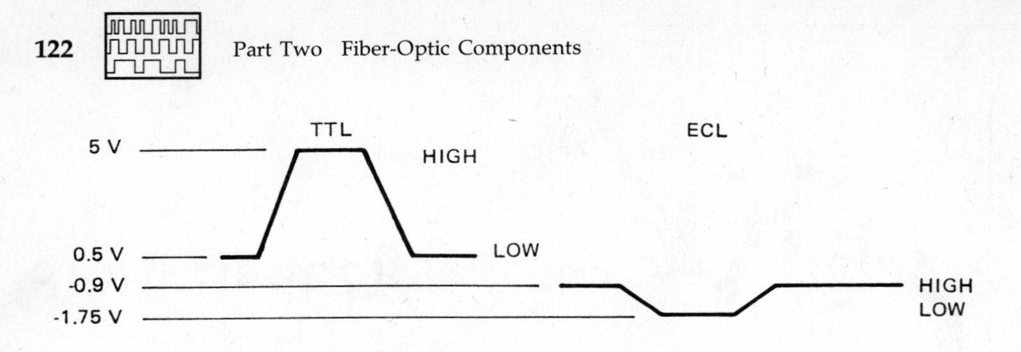

FIGURE 10–2 TTL and ECL signal levels

logic (ECL). ECL circuits use −1.75 V for low and −0.9 V for high. Figure 10–2 shows the logic levels for TTL and ECL circuits.

Normally, TTL and ECL logics cannot be intermixed; however, special chips allow signals to be converted from one logic level to another. Both TTL and ECL are used in fiber-optic transmitters and receivers. TTL is the more common; ECL is usually found only in high-speed systems—those above 50 or 100 Mbps.

Another very popular logic family is complementary metal-oxide semiconductor (CMOS). It is rapidly becoming a popular replacement for TTL because of its very low power consumption. Many CMOS circuits use the same voltage levels as TTL for their signals.

The drive circuit of the transmitter must accept these logic levels. It then provides the output current to drive the source. For example, it may convert the 0.5 V and 5 V of TTL into 0 mA and 50 mA to turn the source on and off.

An additional function of the transmitter is to produce the proper modulation code.

MODULATION CODES

A string of digital highs and lows is often not suitable for transmission over any appreciable distance. Whereas digital circuits use simple high and low pulses to represent 1s and 0s of binary data, a more complex format is often used to transmit digital signals between electronic systems. A modulation code is a method of encoding digital data for transmission.

Earlier, we saw that the pulse represents a binary 1, and the absence of a pulse represents a binary 0. In a TTL system, a 5-V high represents a 1; a 0.5-V low represents a 0. Thus, there is a one-to-one correspondence between a high or low voltage and a binary 1 or 0. With the modulation codes described here, this correspondence does not always exist.

Figure 10–3 shows several popular modulation codes. Each bit of data must occur with its bit period, which is defined by the clock. The *clock* is a steady string of pulses that provides basic system timing. As shown in the table in Figure 10–3, some codes are self-clocking, and others are not. A self-clocking code means that

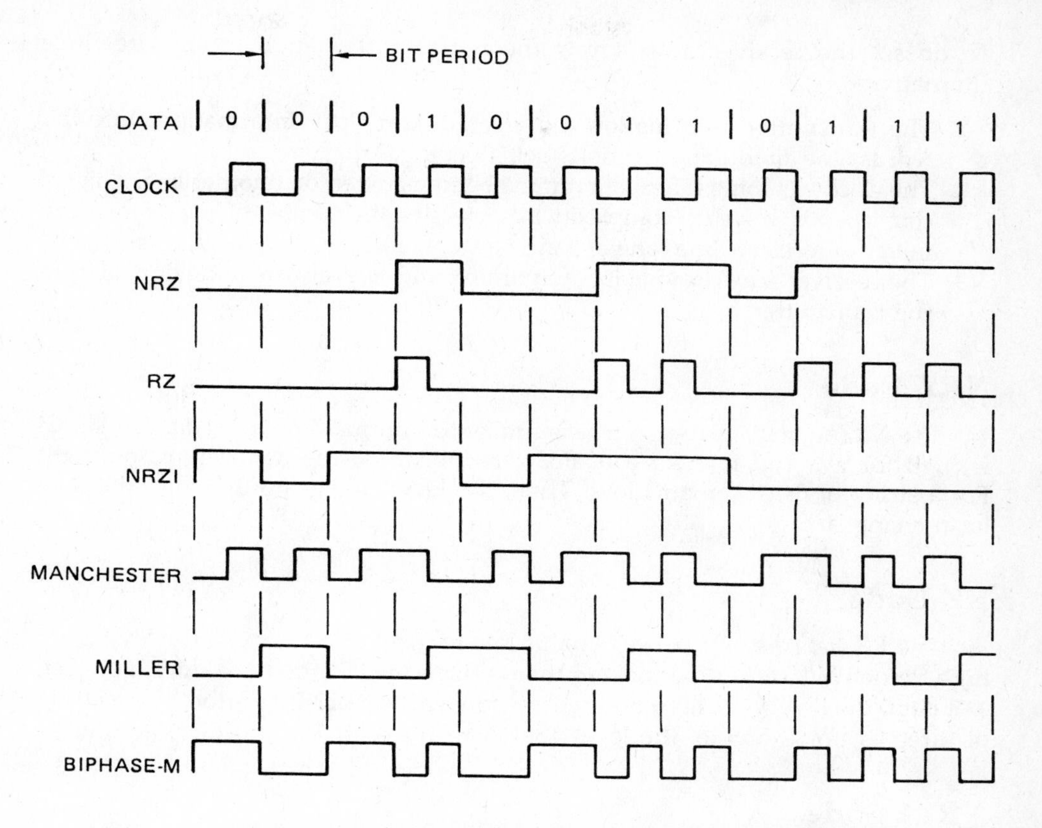

FORMAT	SYMBOLS PER BIT	SELF-CLOCKING	DUTY FACTOR RANGE (%)
NRZ	1	NO	0 - 100
RZ	2	NO	0 - 50
NRZI	1	NO	0 - 100
MANCHESTER (BIPHASE-L)	2	YES	50
MILLER	1	YES	33 - 67
BIPHASE-M (BIFREQUENCY)	2	YES	50

FIGURE 10–3 Modulation codes

the clock information is contained within the code. In a non-self-clocking system, this timing information is not present.

Clock information is important to the receiver. One purpose of a receiver is to rebuild signals to their original state as they were presented to the transmitter.

To do so, the receiver must know the timing information. There are three alternatives:

1. The transmitted information can also contain clock information; in other words, the modulation code is self-clocking.
2. The clock or timing information must be transmitted on another line. This, of course, adds to system complexity by increasing the number of lines from transmitter to receiver. In a long-distance system, such costs may be considerable.
3. The receiver may provide its own timing and not rely on clock signals from the transmitter.

NRZ Code

The *NRZ (nonreturn-to-zero) code* is similar to ''normal'' digital data. The signal is high for a 1 and low for a 0. For a string of 1s, the signal remains high. For a string of 0s, it remains low. Thus, the level changes only when the data level changes.

RZ Code

The *RZ (return-to-zero) code* remains low for 0s. For a binary 1, the level goes high for one half of a bit period and then returns low for the remainder. For each 1 of data, the level goes high and returns low within each bit period. For a string of three 1s, for example, the level goes high for each 1 and returns to low.

NRZI Code

In an *NRZI (nonreturn-to-zero, inverted) code,* a 0 is represented by a change in level, and a 1 is represented by no change in level. Thus, the level will go from high to low or from low to high for each 0. It will remain at its present level for each 1. An important thing to notice here is that there is no firm relationship between 1s and 0s of data and the highs and lows of the code. A binary 1 can be represented by either a high or a low, as can a binary 0.

Manchester Code

A *Manchester code* uses a level transition in the middle of each bit period. For a binary 1, the first half of the period is high, and the second half is low. For a binary 0, the first half is low, and the second half is high.

Miller Code

In the *Miller code,* each 1 is encoded by a level transition in the middle of the bit period. A 0 is represented either by no change in level following a 1 or by a change at the beginning of the bit period following a 0.

Biphase-M Code

In the *bi-phase-M code,* each bit period beings with a change of level. For a 1, an additional transition occurs in midperiod. For a 0, no additional change occurs. Thus, a 1 is at both high and low during the bit period. A 0 is either high or low, but not both, during the entire bit period.

4B/5B and 4B/8B Encoding

Most local area networks use Manchester encoding. Many fiber-optic LANs do not. One reason is that Manchester encoding requires a clock rate twice that of the data rate. A 100-Mbit/s network requires a 200 MHz clock. NRZI data transmissions have no transitions when all zeros are present, which eliminates self-clocking. The receiver cannot properly synchronize its operation with the incoming stream of data.

Many high-speed fiber-optic systems use a group-encoding scheme in which the data bits are encoded into a data word of longer bit length. For example, 4B/5B encoding means that every four data bits are encoded for transmission into a 5-bit code word. The receiver then decodes the 5-bit word to extract the 4 bits. This scheme guarantees that the data never have more than three consecutive zeroes. Such encoding requires less bandwidth than Manchester encoding, increasing the transmission-speed requirements only 20%. FDDI uses this scheme.

Another scheme used in Fiber Channel and in the IBM ESCON system is 4B/8B, which encodes either 4 bits into a 5-bit word or 8 bits into a 10-bit word. FDDI, ESCON, and Fiber Channel are discussed in detail in Chapter 15.

DATA RATE AND SIGNAL RATE

A close inspection of the modulation codes in Figure 10–3 shows an important aspect of signal transmission. The highs and lows may be changing faster than the 1s and 0s of binary data. In the figure, there are 12 data bits (000100110111). The modulation codes, however, use more symbols to represent these bits. A symbol is a high or a low pulse in the modulation code. The Manchester code uses two symbols, both a high and a low, for each binary bit. It always requires twice as many symbols as there are bits to be transmitted.

When we describe the speed of a system, it becomes important to distinguish between data rate and signal rate. *Data rate* is the number of *data bits* transmitted in bits per second. A system may operate at 10 Mbps. The *signal* rate (or speed) is the number of *symbols* transmitted per second. Signal speed is expressed in baud. The signal speed (or baud rate) and the data rate (or bit rate) may or may not be the same, depending on the modulation code used.

For NRZ data, which uses one symbol per bit, the rates are the same. A 10 Mbps requires a 10 Mbaud capacity. For Manchester-encoded data, which uses two symbols per bit, the baud rate is twice that of the bit rate. To transmit a

10-Mbps data stream requires a transmission link having a bandwidth of 20 Mbaud. The baud rate, or number of symbols to be transmitted, is the true indication of a system's signaling speed. This fact is true for all transmission systems, fiber optic or otherwise.

DUTY CYCLE

Duty cycle refers to the ratio of high to low symbols in the encoded string. A duty cycle of 0 means that all symbols are low. A duty cycle of 100 means that all symbols are high. A duty cycle of 50 means that there are an equal number of highs and lows. Duty cycle expresses the relationship between peak and average power levels arriving at the receiver. For a duty cycle of 50%, the average power is half the peak power. Above a 50% duty cycle, the average power increases in relation to peak power. Below 50%, it declines. Duty cycle becomes important in the receiver, and we will delay further discussion until later in this chapter.

TRANSMITTER OUTPUT POWER

The output power of the transmitter is of great concern. The power of many transmitters is specified for the power coupled into a given fiber. Thus for a given transmitter design, the typical output power might be as follows:

Fiber	Power
50/125 μm, 0.21 NA	–16 dBm
62.5/125 μm, 0.275 NA	–12 dBm
85/125 μm, 0.26 NA	–10 dBm
100/140 μm, 0.30 NA	–8 dBm

The power coupled increases with core diameter and NA. The tradeoff, of course, is that the smaller cores and NA usually indicate fibers with lower attenuation and higher bandwidths.

If the transmitter power output is not specified as listed, the power coupled into the fiber must be calculated. To do so, we must know the source output power, output diameter and NA, fiber core diameter and NA, and the expected connector loss. We will look at an example of this calculation in the next chapter.

BASIC RECEIVER CONCEPTS

Figure 10–4 shows a basic diagram of a receiver. It contains the detector, amplifier, and output section. The amplifier amplifies the attenuated signal from the detector. The output section can perform many functions:

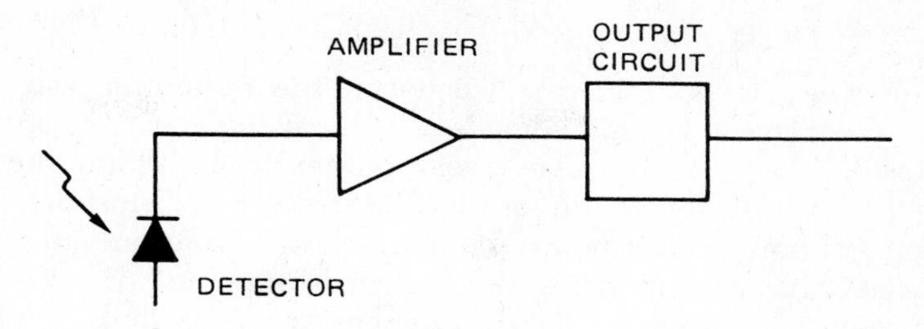

FIGURE 10–4 Basic receiver block diagram

- Separation of the clock and data
- Pulse reshaping and retiming
- Level shifting to ensure compatibility—TTL, ECL, and so forth—with the external circuits.
- Gain control to maintain constant amplification levels in response to variations in received optical power and variations in receiver operation from temperature or voltage changes.

Because the receiver deals with the highly attenuated light signals, it can be considered the principal component around which the design or selection of a fiber-optic system revolves. It is in the photodetector and first stage of amplification that the signal being transmitted is at its weakest and most distorted. It is reasonable to say that this is the central part of the link. Decisions affecting other parts of the link are decided with the receiver in mind. Decisions about the modulation codes in the transmitter are decided, at least in part, by the requirements of the receiver.

Receiver Sensitivity

Receiver sensitivity specifies the weakest optical signal that can be received. The minimum signal that can be received depends on the noise floor of the receiver front end. The last chapter described the limitations on the minimum optical power the detector can receive. The minimum level for the receiver is basically the same, except that the influence of the amplifier noise must also be included. From the noise floor, the lowest *practical* optical power to the receiver depends on the SNR or BER requirements of the system.

Sensitivity can be expressed in microwatts or dBm. Expressing a sensitivity as 1 μW or -30 dBm is saying the same thing. With a packaged receiver, the sensitivity may be specified as an absolute minimum (the noise floor) or with respect to a given level of performance such as BER level.

Dynamic Range

Dynamic range is the difference between the minimum and maximum acceptable power levels. The minimum level is set by the sensitivity and is limited by the detector. The maximum level is set by either the detector or the amplifier. Power levels above the maximum saturate the receiver or distort the signal. The received optical power must be maintained below this maximum.

If a receiver has a minimum optical power requirement of −30 dBm and a maximum requirement of −10 dBm, its dynamic range is 20 dB. It can receive optical inputs between 1 and 100 μW.

Figure 10–5 shows the relationship between BER, data rate, and received power for a typical receiver. As we would expect, BER improves as the received power becomes larger. On the other hand, the required power for a given BER increases significantly as the data rate becomes higher. A BER of 10^{-9} requires an optical input of about −35 dBm (310 nW) for a 25-Mbps data rate and about −32 dBm (620 nW) for a 60-Mbps rate. The received power must be 3 dB, or about 50%, greater for the higher data rate.

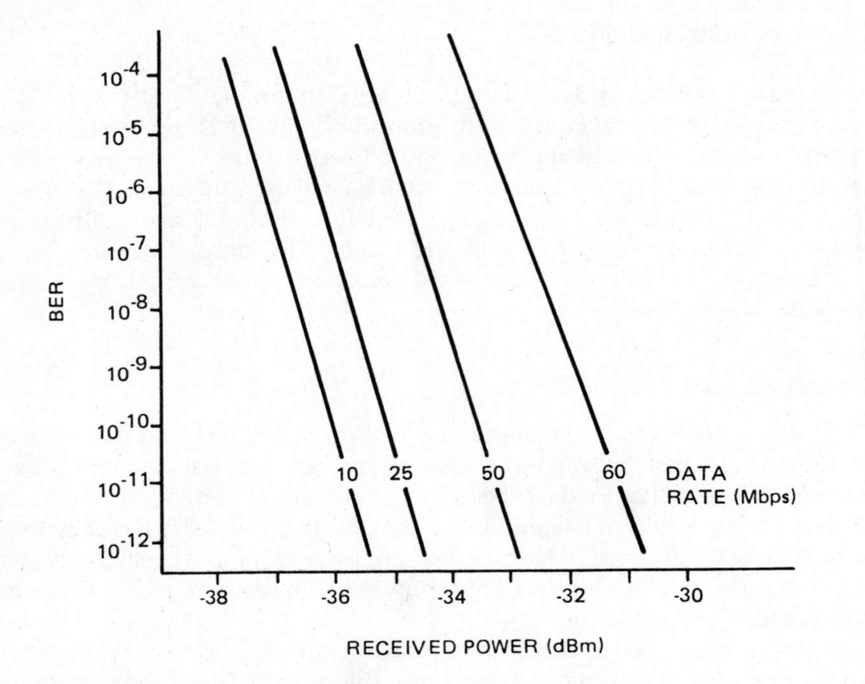

FIGURE 10–5 Relation between received power, BER, and data rate for a typical optical receiver

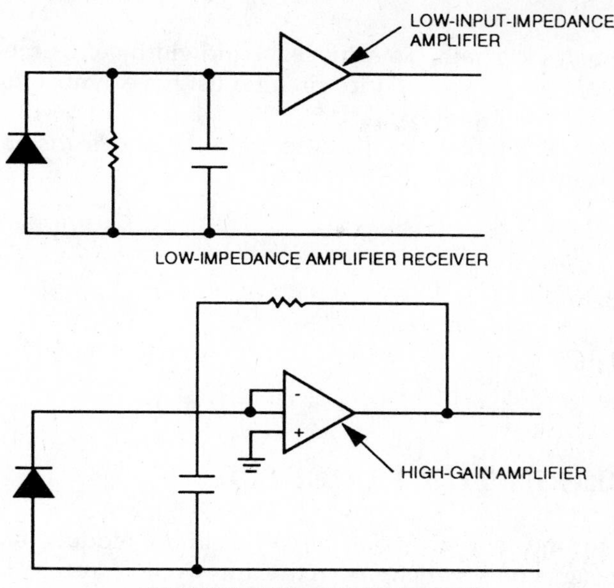

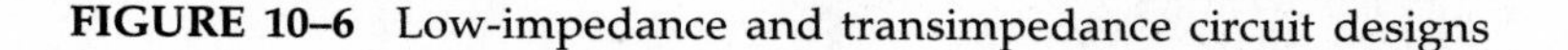

FIGURE 10–6 Low-impedance and transimpedance circuit designs

AMPLIFIER

Two classical designs used in fiber-optic receivers use a low-impedance amplifier and a transimpedance amplifier. Examples are shown in Figure 10–6.

The bandwidth of the low-impedance amplifier is determined by the *RC* time constant of the circuit:

$$\text{BW} = \frac{1}{2\pi RC}$$

This equation is the same as for a photodiode whose response time is limited by the time constant. From the discussion of noise and response time for a photodiode, recall that bandwidth and resistor noise decrease with increasing resistance. Therefore, receiver sensitivity can be increased by increasing the load resistor but at the expense of bandwidth. The resulting high-impedance amplifier has a much smaller bandwidth but greater sensitivity.

The bandwidth of a receiver with a transimpedance amplifier is affected by the gain of the amplifier:

$$\text{BW} = \frac{g}{2\pi RC}$$

where g is the open-loop gain. To achieve a bandwidth comparable to the low-impedance circuit, the transimpedance circuit must have a much lower value for the resistor across the amplifier.

Representative sensitivities for the different receiver designs operating at 100 Mbps and a BER of 10^{-9} are

- low impedance
  −33 dBm
- high impedance
  −40 dBm
- transimpedance
  −36 dBm

DUTY CYCLE IN THE RECEIVER

The reason for the concern about duty cycle in the modulation codes is that some receiver designs put restrictions on the duty cycle. A receiver distinguishes between high and low pulses by maintaining a reference threshold level. A signal level above the threshold is seen as a high or 1; a signal level below the threshold is seen as a low or 0.

In a restricted-duty-circle receiver, a duty cycle far removed from 50% will increase the chances of a level being misinterpreted. A high level is seen as a low level, or a low level is misinterpreted as a high level. As the duty cycle departs from the 50% ideal, the misinterpretations of signal levels degrade the BER.

Figure 10–7 shows why such errors occur in a receiver with a restricted duty cycle receiving NRZ code. The receiver sets the threshold level based on the

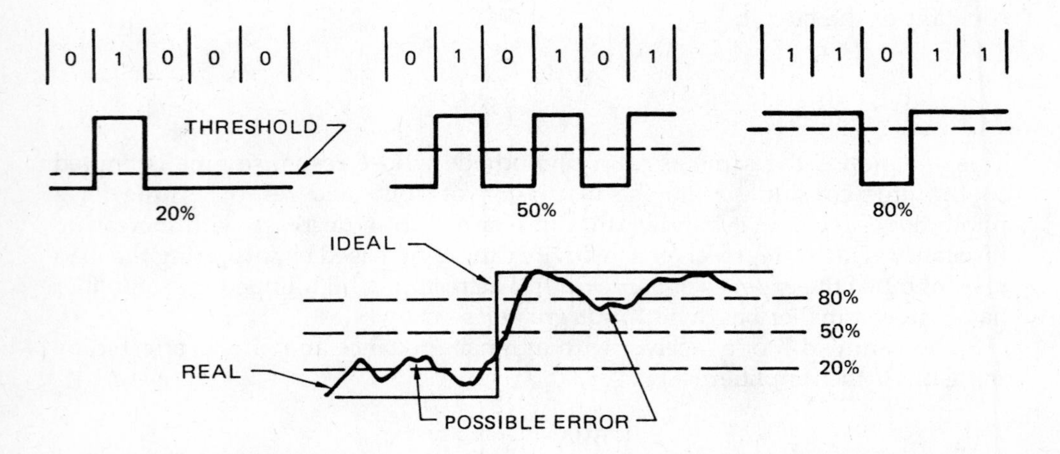

FIGURE 10–7 Errors and duty cycle in a restricted-duty-cycle receiver

average optical power being received. The top part of the figure shows how the threshold level depends on the duty cycle. For a 50% duty cycle, the threshold is set midway between the high and low signal levels, since the average power is midway between signal levels. The difference between the threshold and a low signal level is the same as the difference between the threshold and the high signal level.

At a 20% duty cycle, where low levels significantly outnumber high levels, the average received power is much lower. The threshold shifts much closer to the low signal level. At an 80% duty cycle, the average received power is much higher, and the threshold shifts closer to the high signal level. This shifting results from the average power being received.

This shifting of threshold level would cause no problems in an ideal, noiseless receiver. But receivers are neither perfect nor noiseless. Signal levels not only vary somewhat, but the signals contain noise. The bottom half of the figure shows how noise distorts the pulse. If the signal level crosses the threshold at the wrong point, it is misinterpreted as the wrong level. A 0 can cross the low threshold at a 20% duty cycle and be seen as a 1. A 1 can cross the threshold at an 80% duty cycle and be seen as a 0. In either case, a bit error results.

There are two ways to get around such errors. The first is to maintain a duty cycle close to 50%. Manchester and biphase-M codes, by definition, always have a 50% duty cycle, so they satisfy the requirement. Their drawback is that they require a channel bandwidth of twice the data rate. They also increase the complexity of the transmitter somewhat.

The second method of avoiding bit errors is to design a receiver that maintains the threshold without drift. The reference threshold is always midway between high and low signal levels. One way to do this is by a dc-coupled receiver, which is designed to operate with arbitrary data streams. The receiver is edge-sensing, meaning that it is sensitive to changes in level and not to the levels themselves. This type of receiver reacts only to pulse transitions.

Since the dc-coupled receiver is easy to design and build, why not use it all the time? The tradeoff is receiver sensitivity. It often requires 6 to 8 dB higher optical power levels. This power penalty must be paid by lower data rates, shorter transmission distances, or lower error requirements. Even so, it may often be the best approach in many applications.

In ac coupling, which is one approach used in restricted-duty-cycle receivers, any dc levels are removed from the signal. Only the time-varying portions are left. This stripping of dc levels is often done capacitively, since a capacitor blocks dc and passes ac. The receivers in Figure 10–6 are shown dc coupled to keep the representation simple. An ac-coupled design would place a capacitor between the photodetector and the amplifier.

TRANSCEIVERS AND REPEATERS

Transceivers and repeaters are two important components in many fiber-optic applications. They are simple enough in concept, and we need only comment briefly on each. A *transceiver* is a transmitter and receiver packaged together to allow both transmission and reception from either station. A *repeater* is a receiver driving a transmitter. The repeater is used to boost signals when the transmission distance is so great that the signal will be too highly attenuated before it reaches the receiver. The repeater accepts the signal, amplifies and reshapes it, and sends it on its way by feeding the rebuilt signal to a transmitter.

One advantage of digital transmission is that it uses regenerative repeaters that not only amplify a signal but reshape it to its original form as well. Any pulse distortions from dispersion or other causes are removed. Analog signals use nonregenerative repeaters, which amplify the signal, including any noise or distortion. Analog signals cannot be easily reshaped, because the repeater does not know what the original signal looked like. For a digital signal, it does know.

TRANSMITTER AND RECEIVER PACKAGING

The physical packaging of a transmitter or receiver can be an important aspect in many applications. It can be a small module placed on a printed circuit board, an independent printed circuit board plugged into equipment, or a stand-alone unit. Part of the choice lies in the complexity of the design, and part in the needs of the application. Figure 10–8 shows board-mounting transmitter and receiver modules.

FIGURE 10–8 Transmitter/receiver modules (Illustration courtesy of AMP Incorporated)

TRANSMITTER AND RECEIVER SPECIFICATIONS

Many of the characteristics for specifying a transmitter or receiver are those of concern with any electronic circuit. These include power supply voltages, storage and operating temperature ranges, required input and output voltage levels (which indicate TTL or ECL compatibility), propagation delays, and so forth. There are also important specifications relating to the transmitter or receiver in a fiber-optic system. We have already become acquainted with each of these.

Transmitter

- Peak output power
- Data rate/bandwidth
- Operating wavelength
- Source spectral width
- Duty-cycle restrictions
- Rise and fall times

Receiver

- Data rate/bandwidth
- Sensitivity
- Dynamic range
- Operating wavelength
- Duty-cycle restrictions

Table 10–1 supplies the specifications for a 125-Mbps fiber-optic transmitter-receiver that meet the requirements of the FDDI application described in Chapter 15.

Parameter	Conditions	Min	Typ	Max	Units
TRANSMITTER SECTION					
Data Rate, NRZ		0		125	Mbps
Optical Output Power	62.5/125, 0.275 NA Fiber	–19		–14	dBm
Center Wavelength		1270	1330	1380	nm
Spectral Width	FWHM		140		nm
Duty Cycle	Data Rate = 0–125 Mbps	0		100	%
Risetime, Output	10–90%	0.6	2	3	ns
Falltime, Output	10–90%	0.6	2	3	ns
Supply Voltage		4.5	5.0	5.72	V
Supply Current				150	mA
Operating Temperature		0	25	70	°C
RECEIVER SECTION					
Data Rate		10		125	Mbps
Sensitivity	62.5/125, 0.275 NA Fiber, Data Rate = 125 Mbps, BER = 2.5×10^{-10}	–33		–14	dBm
Wavelength		1270		1390	nm
Duty Cycle		30	50	70	%
Risetime, Output	20–80%	0.5		3	ns
Falltime, Output	20–80%	0.5		3	ns
Supply Voltage		4.5	5	5.5	V
Supply Current				150	mA
Operating Temperature		0	25	70	°C

TABLE 10–1 FDDI transceiver specifications

SUMMARY

- A transmitter includes the source and its drive circuit.
- A receiver includes the detector, amplification circuits, and output circuits.
- Modulation codes are the formats by which digital bits are encoded for transmission.
- For some modulation codes, data rate and baud rate are the same. For others, they are not the same.
- Some modulation codes are self-clocking.
- Duty cycle expresses the ratio between high levels and low levels.
- The receiver sensitivity sets the lowest acceptable optical power that can be received.
- The dynamic range of the receiver is the difference between the minimum and maximum optical power that a receiver can handle.

REVIEW QUESTIONS

1. Sketch a block diagram of a simple transmitter.
2. A Manchester-encoded transmission has a data rate of 200 Mbps. What is its baud rate?
3. Name two examples of self-clocking modulation codes.
4. Sketch an NRZ pattern for 1111001. Be sure to show each bit period.
5. Will a typical transmitter couple more power into a 50/125-μm fiber or an 85/125-μm fiber? Why?
6. A receiver has a sensitivity of –30 dBm and a maximum receivable power of –10 dBm. What is its dynamic range?
7. Name two of the commonest types of receiver circuits.
8. Sketch a block diagram of a repeater.
9. If a receiver has a sensitivity of –28 dBm, what is the minimum optical power it can receive?

CHAPTER 11

Connectors and Splices

Interconnection of the various components of a fiber-optic system is a vital part of system performance. Connection by splices and connectors couples light from one component to another with as little loss of optical power as possible. Throughout a link, a fiber must be connected to sources, detectors, and other fibers. This chapter describes the basic considerations involved in fiber-optic interconnections and describes several examples of actual connectors and splices. To simplify the discussion, we emphasize connecting one fiber to another.

By popular usage, a *connector* is a disconnectable device used to connect a fiber to a source, detector, or another fiber. It is designed to be easily connected and disconnected many times. A *splice* is a device used to connect one fiber to another permanently. Even so, some vendors do offer disconnectable splices that are not permanent and that can be disconnected for repairs or rearrangement of circuits.

The requirements for a fiber-optic connection and a wire connection are very different. Two copper conductors can be joined directly by solder or by connectors that have been crimped or soldered to the wires. The purpose is to create intimate contact between the mated halves to maintain a low-resistance path across the junction. Connectors are simple, easy to attach, reliable, and essentially free of loss.

The key to a fiber-optic interconnection is precise alignment of the mated fiber cores (or spots in a single-mode fibers) so that nearly all the light is coupled from one fiber across the junction into the other fiber. Contact between the fibers is not even mandatory. The demands of precise alignment on small fibers create a challenge to the designer of the connector of splice.

THE NEED FOR CONNECTORS AND SPLICES

Connectors and splices are required for many reasons. Although these reasons may be fairly obvious since they mirror the same needs with metallic conductors, we will briefly review them here.

In a long link, fibers may have to be spliced end to end because cable manufacturers offer cables in limited lengths—typically 1 to 6 km. If the cable comes in 6-km lengths, a 30-km span requires four splices (as well as connections to the transmitter and receiver). In other cases, it may not be practical to even pull a 6-km length of cable through ducts during installation. More moderate lengths may be easier to install.

Connectors or splices may be required at building entrances, wiring closets, couplers, and other intermediate points between transmitter and receiver. These allow, for example, transitions between outdoor and indoor cables, rearrangement of circuits, and the division of optical power from one fiber into several fibers.

Finally, fibers must be connected to the source in the transmitter and the detector in the receivers.

Dividing a fiber-optic system into several subsystems connected together by splices and connectors also simplifies system selection, installation, and maintenance. Components may be selected from different vendors. They may be installed by different vendors or contractors. For example, a building contractor can wire a building with fiber-optic cable, a manufacturer of computer terminals can build fiber-optic transmitters and receivers in the equipment, and the telephone company can bring a fiber-optic telephone line into the building. All these various parts are then linked together with connectors. Maintenance of the system is simplified when a faulty or outdated part can be disconnected and a new or updated part installed. Faster transmitters and receivers, for instance, can be installed without disturbing the fiber.

CONNECTOR REQUIREMENTS

The following is a list of desirable features for a fiber-optic connector or splice:

- *Low loss:* The connector or splice should cause little loss of optical power across the junction.
- *Easy installation:* The connector or splice should be easily and rapidly installed without need for extensive special tools or training.
- *Repeatability:* A connector should be able to be connected and disconnected many times without changes in loss.
- *Consistency:* There should be no variation in loss; loss should be consistent whenever a connector is applied to a fiber.
- *Economical:* The connector or splice should be inexpensive, both in itself and in special application tooling.

It can be very difficult to design a connector that meets all the requirements. A low-loss connector may be more expensive than a high-loss connector, or it may require relatively expensive application tooling. The lowest losses are desirable, but the other factors clearly influence the selection of the connector or splice.

In general, loss requirements for splices and connectors are as follows:

- 0.2 dB or less for telecommunication and other splices in long-haul systems.
- 0.3 to 1 dB for connectors used in intrabuilding systems, such as local area networks or automated factories.
- 1 to 3 dB for connectors and splices used in applications where such higher losses are acceptable and low cost becomes more important than low loss. Such applications often use plastic fibers.

CAUSES OF LOSS IN AN INTERCONNECTION

Three different types of factors cause loss in fiber-optic interconnections:

1. Intrinsic or fiber-related factors are those caused by variations in the fiber itself.
2. Extrinsic or connector-related factors are those contributed by the connector itself.
3. System factors are those contributed by the system.

INTRINSIC FACTORS

In joining two fibers to each other, we would like to assume that the two fibers are identical. Usually, however, they are not. The fiber manufacturing process allows fibers to be made only within certain tolerances. From the nominal specification, a fiber will vary within stated limits. Figure 11–1 shows schematically the most important variations in tolerances that cause loss. The following paragraphs discuss them in greater detail.

NA-mismatch loss occurs when the NA of the transmitting fiber is larger than that of the receiving fiber. *Core-diameter-mismatch loss* occurs when the core or diameter of the transmitting fiber is larger than that of the receiving fiber. *Cladding-diameter-mismatch loss* occurs when the claddings of the two fibers differ, since the cores will no longer align.

Concentricity loss occurs because the core may not be perfectly centered in the cladding. Ideally, the geometric axes of the core and cladding should coincide. The concentricity tolerance is the distance between the core center and the cladding center.

Ellipticity (or *ovality*) *loss* occurs because the core or cladding may be elliptical rather than circular. Notice that the alignment of the two elliptical cores will vary, depending on how the fibers are brought together. At one joining, they may be at the extremes of ellipticity so that maximum loss occurs. At the next joining, one fiber may become rotated in relation to the other, so their elliptical axes are the same. No loss then occurs. The ellipticity or ovality tolerance of the core and cladding equals the minimum diameter divided by the maximum diameter.

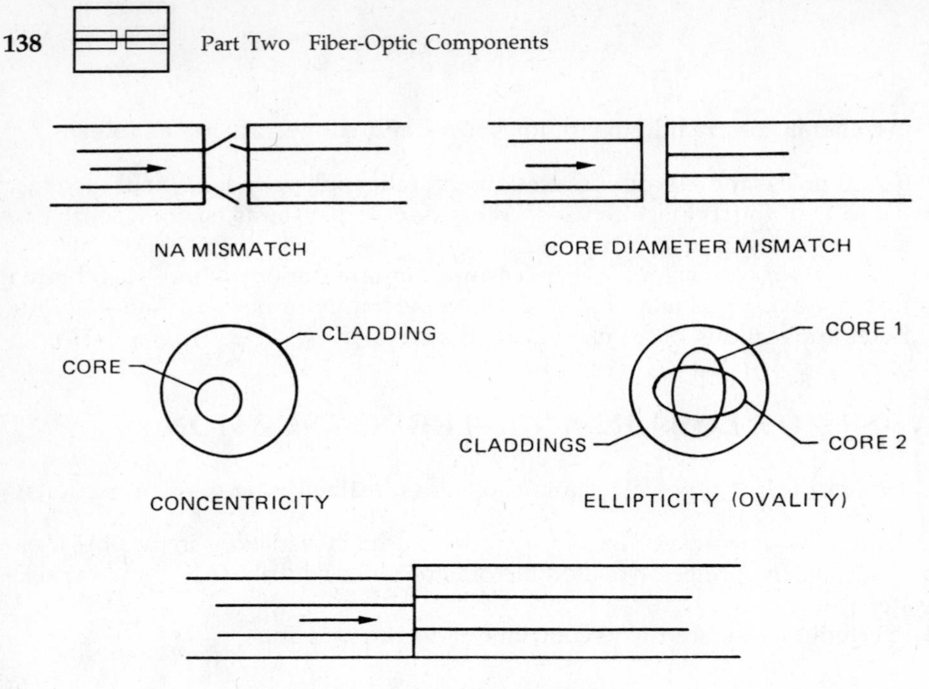

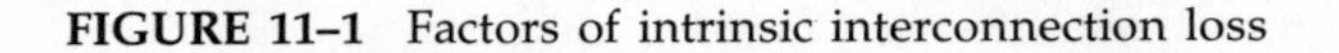

FIGURE 11–1 Factors of intrinsic interconnection loss

These variations exist in any fiber. The fiber manufacturer controls these variations by manufacturing fiber to tight, exacting tolerances. In the past few years, manufacturing techniques have improved sufficiently that fiber tolerances are much tighter. For example, a 125-μm-diameter fiber once had a tolerance of ± 5 μm, so the actual fiber diameter was in a range of 120 to 130 μm. Two fibers at the extreme of tolerance mismatch present a loss of 0.6 dB. Today, the normal tolerance is ± 2 μm, resulting in a range of 123 to 127 μm and a maximum loss of 0.28 dB. A tolerance of ± 1 μm reduces the loss to 0.1 dB.

Remember that these losses are maximums and will not be experienced in most cases since the probability of joining two fibers at the extreme diameter mismatches is small.

Table 11–1 shows typical tolerances for today's off-the-shelf fibers.

Type of Variation	Tolerance
Core diameter (62.5 μm)	± 3 μm
Cladding diameter (125 μm)	± 2 μm
NA (0.275)	± 0.015
Concentricity	≤ 3 μm
Core Ovality	≥ 0.98
Cladding Ovality	≥ 0.98

TABLE 11–1 Typical tolerances that influence intrinsic loss

EXTRINSIC FACTORS

Connectors and splices also contribute loss to the joint. When two fibers are not perfectly aligned on their center axes, loss occurs even if there is no intrinsic variation in the fibers. The loss results from the difficulty of manufacturing a device to the exacting tolerances required. As we will see, many different alignment mechanisms have been devised for joining two fibers.

The four main causes of loss that a connector or splice must control are

1. Lateral displacement
2. End separation
3. Angular misalignment
4. Surface roughness

The following discussion assumes perfect fibers, without tolerance variation. The curves presented are general ones for multimode fibers. As discussed later in this chapter, modal conditions in the fiber also influence loss across the junction.

Lateral Displacement

A connector should align the fibers on their center axes. When one fiber's axis does not coincide with that of the other, loss occurs. As shown in the graph in Figure 11–2, the amount of displacement depends on the ratio of the lateral offset to the fiber diameter. Thus the acceptable offset becomes less as the fiber diameter becomes smaller. A displacement of 10% of the core axis yields a loss of about 0.5 dB. For a fiber with a 50-μm core, a 10% displacement is 5 μm, which, in turn, means that each connector half must contribute only a 2.5 μm.

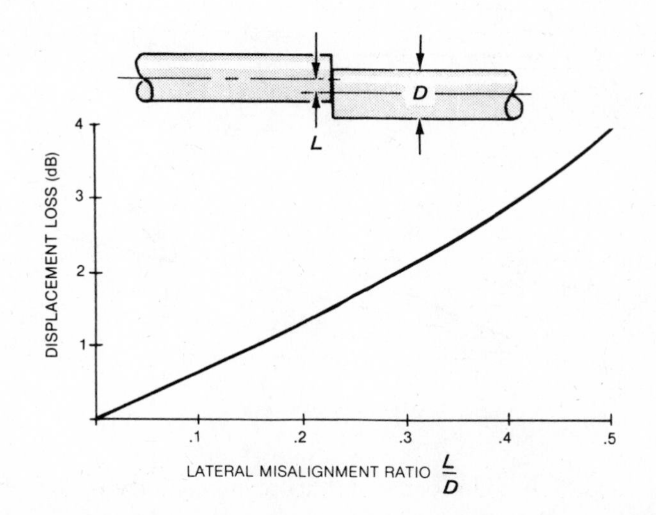

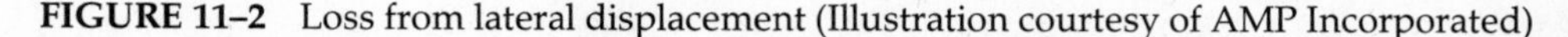
FIGURE 11–2 Loss from lateral displacement (Illustration courtesy of AMP Incorporated)

These dimensions would double for a fiber with a 100-μm core: a 10% displacement would allow 10 μm or 5 μm for each connector half. Obviously, then, control of lateral displacement becomes more difficult for smaller-diameter fibers. Connector manufacturers attempt to limit displacement to less than 5% of the core diameter.

End Separation

Two fibers separated by a small gap experience two types of loss, as shown in Figure 11–3. The first is *Fresnel reflection loss,* which is caused by the difference in refractive indices of the two fibers and the intervening gap, usually air. Fresnel reflection occurs at both the exit from the first fiber and at the entrance to the second fiber. For glass fibers separated by air, loss from Fresnel reflections is about 0.34 dB. Fresnel losses can be greatly reduced by using an index-matching fluid between the fibers. *Index-matching fluid* is an optically transparent liquid or gel having a refractive index the same as or very near that of the fibers.

The second type of loss for multimode fibers results from the failure of high-order modes to cross the gap and enter the core of the second fiber. High-order rays exiting the first fiber will miss the acceptance cone of the second fiber. Light exiting the first fiber spreads out conically. The degree of separation loss, then, depends on fiber NA. A fiber with a high NA cannot tolerate as much separation to maintain the same loss as can a fiber with a lower NA.

Ideally, the fibers should butt to eliminate loss from end separation. In most splices, the fibers do butt. In a separable connector, a very small gap is sometimes

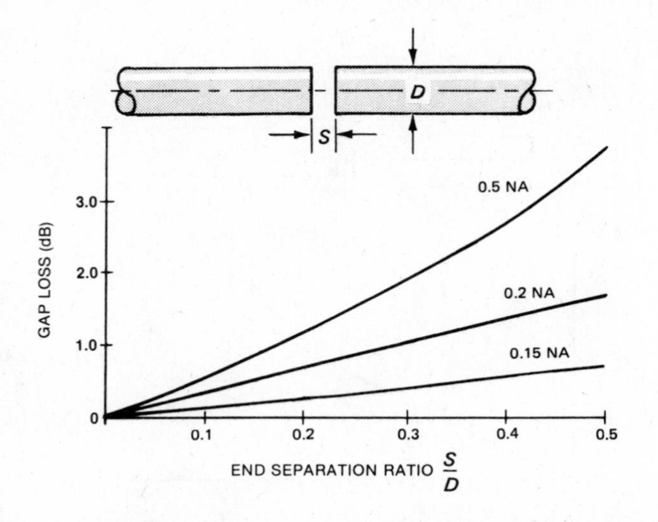

FIGURE 11–3 Loss from end separation (Illustration courtesy of AMP Incorporated)

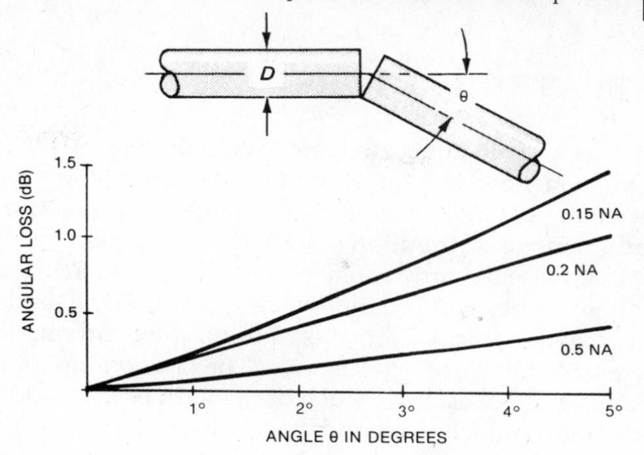

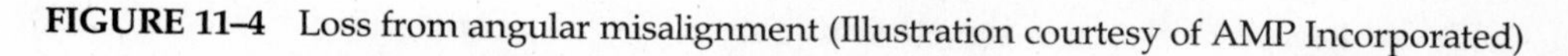
FIGURE 11–4 Loss from angular misalignment (Illustration courtesy of AMP Incorporated)

desirable to prevent the ends from rubbing together and causing abrasion during mating. Fibers brought together with too much force, as may be done by overtightening coupling nuts on a connector, may even be fractured. Therefore, some connectors are designed to maintain a very small gap between fibers, and others use spring pressure to bring the fiber ends gently together without the danger of damage. Physical contact of the fibers is often desirable to control return reflections, which are discussed later in this chapter.

Angular Misalignment

The ends of mated fibers should be perpendicular to the fiber axes and perpendicular to each other during engagement. The losses graphed in Figure 11–4 result when one fiber is cocked in relation to the other. Again, the degree of loss also depends on the fiber NA. Notice, however, that the influence of NA is opposite that for end separation. Here larger NA can tolerate greater angular misalignment to maintain the same loss as for a lower NA.

A properly applied connector controls angular misalignment rather easily, so its loss contribution is not as great as with lateral displacement. This control includes having the fiber face perpendicular to the fiber axis, which is a function of proper fiber cleaving or polishing, and preventing the connector from cocking during engagement.

Surface Finish

The fiber face must be smooth and free of defects such as hackles, burrs, and fractures. Irregularities from a rough surface disrupt the geometrical patterns of light rays and deflect them so they will not enter the second fiber.

SYSTEM-RELATED FACTORS

The loss at a fiber-to-fiber joint depends not only on the losses contributed by the connector and fiber but on system-related factors as well. Chapter 6 discussed how modal conditions in a fiber change with length until the fiber reaches equilibrium mode distribution (EMD). Initially, a fiber may be overfilled or fully filled, with light being carried both in the cladding and in high-order modes. Over distance, these modes will be stripped away. At EMD, a graded-index fiber has a reduced NA and a reduced active area of the core carrying the light.

Consider a connection close to the source. The fiber on the transmitting side of the connection may be overfilled. Much of the light in the cladding and high-order modes will not enter the second fiber, although it was present at the junction. This same light, however, would not have been present in the fiber at EMD, so it would also not have been lost at the interconnection point.

Next consider the receiving side of the fiber. Some of the light will spill over the junction into cladding and high-order modes. If we were to measure the power from a short length of fiber, these modes would still be present. But these modes will be lost over distance, so their presence is misleading.

Similar effects will be seen if the connection point is far from the source, where the fiber has reached EMD. Since the active area of a graded-index fiber has been reduced, lateral misalignment will not affect loss as much, particularly if the receiving fiber is short. Again, light will couple into cladding and high-order modes. These modes will be lost in a long receiving fiber.

Thus, the performance of a connector depends on modal conditions and the connector's position in the system (since modal conditions vary along the length of the fiber). In evaluating a fiber-optic connector or splice, we must know conditions on both the launch (transmitter) side and the receive (receiver) side of the connection. Four different conditions exist:

1. Short launch, short receive
2. Short launch, long receive
3. Long launch, short receive
4. Long launch, long receive

With all other things held constant, connector performance depends on the launch and receive conditions. For a series of tests done on one connector, for example, losses under long-launch conditions were in the 0.4- to 0.5-dB range. Under short-launch conditions, the losses were in the 1.3- to 1.4-dB range. Thus, a difference of nearly 1 dB occurred simply as a result of the launch conditions used in the test.

INSERTION LOSS

Insertion loss is the method for specifying the performance of a connector or splice. Power through a length of fiber is measured. Next, the fiber is cut in the center and the connector or splice is applied. Power at the end of the fiber is again measured. Insertion loss is

$$\text{loss}_{IL} = 10 \log_{10} \left(\frac{P_1}{P_2} \right)$$

where P_2 is the initial measured power and P_1 is the measured power after the connector is applied.

This testing method, by using a single fiber, minimizes the influence of fiber variation on loss. By rejoining the cut fiber, we make the mated halves essentially identical. The only fiber variations present are eccentricity and ellipticity. NA and diameter variations are eliminated. The purpose is to evaluate connector performance independently of fiber-related variations.

Launch and receive conditions still affect the measured insertion loss. Therefore, these conditions must be known when comparing insertion loss specifications for various connectors. Differences in loss specifications may result from differences in test conditions. In addition, the application in which the connector will be used must also be considered. It is best to evaluate a connector that has been tested under conditions similar to those in which it will be used.

ADDITIONAL LOSSES IN AN INTERCONNECTION

The discussion of intrinsic fiber loss stated that loss occurs when the transmitting fiber has a core diameter or NA smaller than the core diameter or NA of the receiving fiber. This discussion assumed the connection was between two fibers of the same type, differing only in variations in tolerance. In this case, these variations may affect the performance of the connector, but the losses from the mismatches are not usually calculated in general loss budgets. If, however, two different types of fibers are connected, diameter- and NA-mismatch losses may be significant and must be accounted for.

When the NA of the transmitting fiber is greater than that of the receiving fiber, the NA-mismatch loss is

$$\text{loss}_{NA} = 10 \log_{10} \left(\frac{NA_r}{NA_t} \right)^2$$

Similarly, when the diameter of the transmitting fiber is greater than that of the receiving fiber, the diameter-mismatch loss is

$$\text{loss}_{dia} = 10 \log_{10} \left(\frac{dia_r}{dia_t} \right)^2$$

Assume, for example, a transmitting fiber with a core diameter of 62.5 μm and an NA of 0.275 and a receiving fiber with a diameter of 50 μm and an NA of 0.20. Such a connection could be experienced in connecting fibers in a local area network. Loss from NA-mismatch is

$$\begin{aligned} \text{loss}_{\text{NA}} &= 10 \log_{10} \left(\frac{\text{NA}_r}{\text{NA}_t} \right)^2 \\ &= 10 \log_{10} \left(\frac{0.20}{0.275} \right)^2 \\ &= -2.8 \text{ dB} \end{aligned}$$

Loss from diameter mismatch is

$$\begin{aligned} \text{loss}_{\text{dia}} &= 10 \log_{10} \left(\frac{\text{dia}_r}{\text{dia}_t} \right)^2 \\ &= 10 \log_{10} \left(\frac{50}{62.5} \right)^2 \\ &= -2.9 \text{ dB} \end{aligned}$$

The total loss is 5.7 dB.

LOSS IN SINGLE-MODE FIBERS

Connectors and splices for single-mode fibers must also provide a high degree of alignment. In many cases, the percentage of misalignment permitted for a single-mode connection is greater than for its multimode counterpart. Because of the small size of the fiber core, however, the actual dimensional tolerances for the connector or splice remain as tight or tighter. For example, a gap of 10 times the core diameter results in a loss of 0.4 dB in a single-mode fiber.

Mismatches in mode field diameter (spot size) are the greatest cause of intrinsic, fiber-related loss. Most manufacturers limit spot size deviations to 10%, which results in attenuation increases of under 0.1 dB in 99% of the cases where fibers from the same manufacturer are connected. Since different manufacturers have different values for mode field diameter, greater loss may be expected when mating connectors from different manufacturers.

RETURN REFLECTION LOSS

You learned earlier that when two fibers are separated by an air gap, optical energy will be reflected back toward the source. This energy is termed *return reflection* or *return loss*. The loss refers to the fact that it is energy representing loss at the fiber-to-fiber juncture. It is that part of the loss that reflects and is

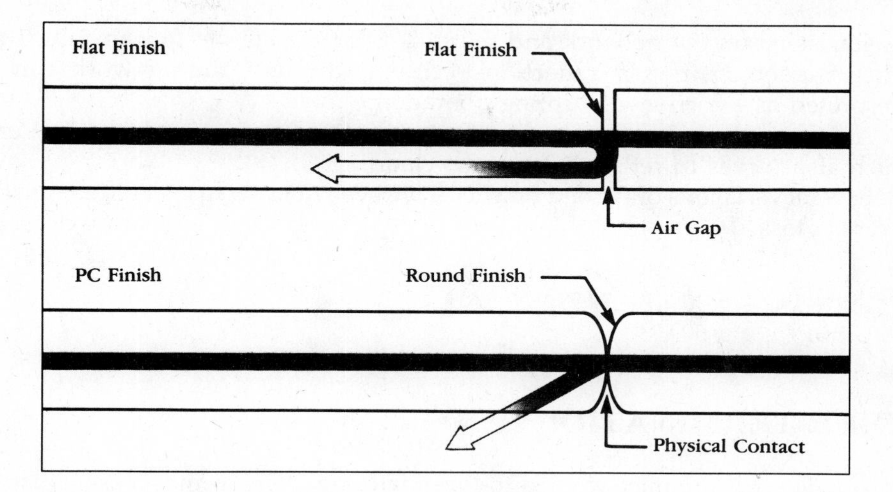

FIGURE 11–5 Return reflections and PC end finish (Illustration courtesy of AMP Incorporated)

propagated in a backwards direction by the fiber. Figure 11–5 shows the idea of return reflections.

In a single-mode interconnection with flat end finish, this loss amounts to –11 dB—that is, the reflected energy is about 11 dB below the incident energy. If 500 μW of energy reach the fiber end, about 40 μW will reflect back toward the source. This level of energy is sufficient to interfere with operation of a laser diode. In single-mode systems, it is especially important to minimize return loss so that proper operation of the source is not disrupted.

Either by assuring fiber-to-fiber contact or eliminating air from the fiber interface, return loss can be lowered to over –30 dB—the 500 μW now reflect only 0.5 μW toward the source. One way to do this is by rounding the fiber end during polishing. Two factors contribute to reduction of return power in a rounded end finish. First, interconnected fiber can now be allowed to touch, which curtails reflection from the mismatches in refractive indices of air and fibers. Only minor reflections occur because of tolerance differences in fiber properties. Second, the rounded end means that light no longer reflects directly back toward the source. It reflects back at an angle and is usually lost from the fiber.

Why not use a flat end finish and butt the fibers? Because achieving two *perfectly* flat, *perfectly* perpendicular ends is difficult. Most likely, one or both of the fibers will have a slight angle, enough to keep the fiber cores from touching. With a rounded finish, fibers always touch on the high side near the light-carrying fiber core.

A rounded PC end finish can find application in multimode systems as well. In some local area networks, for example, the system relies on dead silence between packets of information. The silence indicates that there are no transmissions on the network and that it is OK for a station to transmit. If not reduced enough, return reflections echoing through the network can be interpreted as a carrier, disrupting operation.

Another approach to increasing return reflection loss is to angle the fiber end, which also serves to reflect light in the cladding.

Several variations of PC end finishes have been devised to give different levels of return loss:

PC:	>30 dB
SuperPC:	>40 dB
UltraPC:	>50 dB

FIBER TERMINATION

A connector or splice is used to terminate the fiber. In most cases, a splice is simpler than a connector. The permanence of splices allows a connection device that has fewer component parts. A fusion splice, in fact, has no parts in its simplest form. A fiber-optic connector or splice must fulfill several functions:

- Align the fibers optically
- Secure the fiber in the connector or splice
- ''Decouple'' the fiber from the cable. By this, we mean that the cable, usually the strength members, is also secured so that a pull or tension on the cable remains on the cable rather than on the more fragile fiber.

Figure 11–6 shows a cross-sectional view of two mated connectors. The connectors are held aligned with each other by a coupling receptacle, which is basically a precision-toleranced bore with an outside that provides a means to fasten the connector. The bore is typically either metal or metal with a plastic insert that provides a resilient compression on the connector.

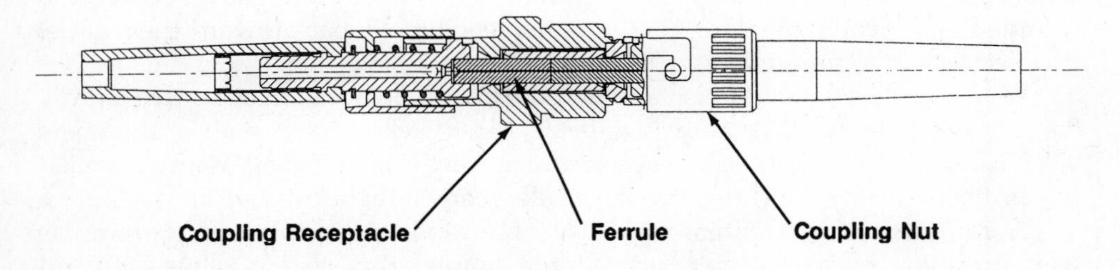

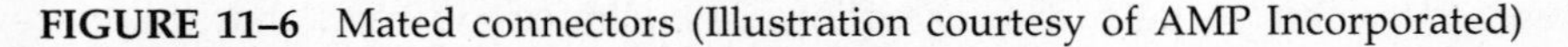
FIGURE 11–6 Mated connectors (Illustration courtesy of AMP Incorporated)

Ferrules

Most connectors use a ferrule to hold the fiber and provide alignment. In one sense, many connectors are simply different ways to build a practical connector around a ferrule.

Ceramic ferrules offer the best performance and are the preferred material for single-mode fibers. Ceramics are strong and have precision, machined fiber bores. In addition, they have excellent thermal and mechanical properties so that performance does not vary due to temperature or environmental fluctuations.

Plastic ferrules offer lower cost but more modest performance. Stainless steel ferrules fall between ceramic and plastic in performance and cost, but they are popular because of their strength—they are less susceptible to breakage or cracking than ceramic.

Two types of ceramic are used for ferrules: alumina oxide and zirconia oxide. Alumina oxide was the first material used: It is hard, inelastic material that allows manufacturers to hold tolerances very precisely. Its coefficient of thermal expansion—how much the material expands when heated and contracts when cooled—is very close to that of glass. The drawback to alumina is that it is more brittle and can break under excess stress. In addition, alumina is much harder to polish, especially in a field installation.

Zirconia oxide is softer than alumina oxide and more resistant to impact. It is still hard enough to hold tolerances nearly as well as alumina oxide, but soft enough to allow faster polishing.

The most popular ferrule size is a 2.5-mm diameter, which has become nearly a standard.

Epoxy and Polish

Fibers are most often epoxied into the connector. Epoxy, because of its curing time, is generally considered to be an undesirable but necessary step in a fiber-optic termination. Epoxy provides good tensile strength to the termination to prevent the fiber from moving within the connector body. It also prevents pistoning (Figure 11–7)—the in/out movement of the fiber within the ferrule. One cause of pistoning is changes in temperature. With changes in temperature, the fiber can expand out of the ferrule or contract within it. If the fiber pistons outward, it extends past the end of the fiber and can be damaged. If the fiber pistons inward, it increases the gap between fibers and thereby the loss. After the epoxy cures, the fiber/connector is polished to a smooth end finish.

Epoxyless Terminations

Even so, epoxyless connectors have been devised. One example (Figure 11–8) uses an internal insert that grips the fiber at the front and rear, providing the stability and tensile strength of epoxy. As the connector is crimped, the insert compresses around the fiber at both the front and the rear.

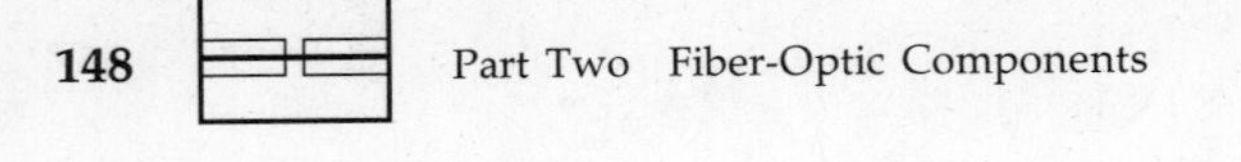

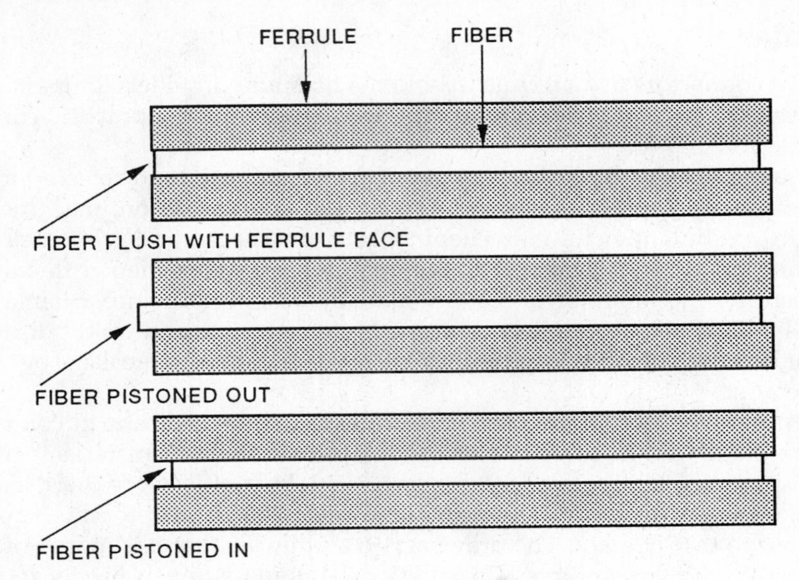

FIGURE 11–7 Fiber pistoning

The front clamp is on the bare fiber. This clamp prevents pistoning. It does not, however, provide the tensile strength to completely prevent movement of the fiber in the connector, and increasing the clamping pressure at this point runs the risk of damaging the fiber. The rear compression adds higher clamping force on the fiber buffer coating to provide the necessary tensile strength. This gripping point will not prevent pistoning during changes in temperature. It is only when both compressions are properly balanced that the fiber is securely held within the connector enough that epoxy is not needed.

Performance is comparable to its epoxy counterparts. The main advantage of an epoxyless connector is speed of assembly. Some users will tolerate a slightly higher loss to achieve a fast, easy termination. Time, after all, is money. A new, large building being wired for fiber may contain thousands of connectors. Speed of installation may be more important than achieving losses much lower than required by the application.

Notice that the epoxyless approach is a technique that is not limited to one connector style.

COMPATIBILITY

A growing trend in electronics is standardization. Standardization implies that a component—anything from a connector to a computer—meets specified

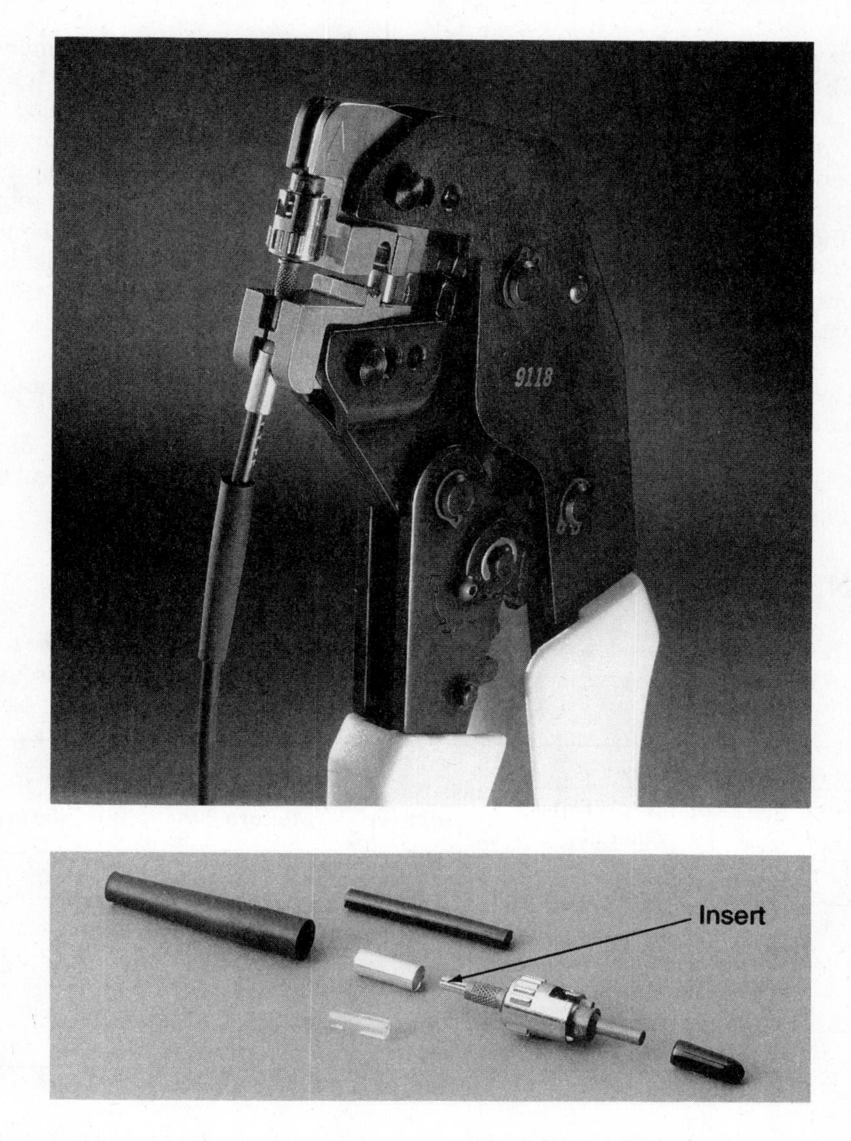

FIGURE 11–8 Expoxyless connector, shown both with component parts and with hand tool, uses an internal insert to grip fiber (Illustration courtesy of AMP Incorporated)

requirements and is interchangeable with other components meeting that standard. The two components are compatible. Whereas some standards are *de facto* ones—they evolve through popularity and not formal standards—most standards are formally created by such industry associations as the Electronics

Industry of America (EIA), the Institute of Electrical and Electronic Engineers (IEEE), or the American National Standards Institute (ANSI).

Compatibility here refers to the need for the connector to be compatible with other connectors or with specifications. For example, large major telecommunication companies have developed connectors for use in their equipment and applications. For many applications, it becomes important for a competitive connector to mate with these. In other cases, specifications describe the type of connector to be used. MIL-C-83522, for instance, defines SMA-style connectors for military applications.

Compatibility exists on several levels. The most basic level is physical compatibility: The connector must meet certain dimensional requirements to allow it to mate with other connectors of the same style.

The next level of compatibility involves performance: insertion loss, durability, temperature range, and so forth.

Finally, standards like MIL-C-83522 define the connector completely: dimensions, performance, materials. Here, compatibility comes to mean identical since the military specification leaves very little room for differences.

CONNECTOR EXAMPLES

The following pages describe some of the most common connectors in use today. Table 11–2 provides a summary of the characteristics of the connectors discussed.

Connector/Splice		Insertion Loss (dB) typ	Single-Mode Return Loss (dB)	Durability (Mating/Unmating Cycles)
Type	Application/Ferrule Material[1]			
FC/PC	SM/MM	0.3	40 typ	1000
ST	SM	0.3	40 typ	1000
	Ceramic MM	0.3		1000
	Stainless Steel MM	0.6		1000
	Plastic MM	0.7		250
SC	SM/MM	0.3	40 typ	1000
FDDI	SM	0.3	35 min	500
	MM	0.5		500
ESCON	MM	0.5		500
SMA	MM	1.5		200
DNP	Plastic Fiber	2		
Finger Splice	SM/MM	0.2	40 typ	NA

1. SM = Single mode; MM = multimode. Materials are ferrule materials. Single-mode connectors all use ceramic ferrules.

TABLE 11–2 Performance characteristics of typical connectors

The late 1980s and early 1990s saw great progress in standardization of connectors. Bear in mind that compatibility does not mean connectors are identical. The great drive in compatibility is the interface: the ability of connectors from different vendors to mate with one another. This still leaves room for different manufacturers to differentiate their products. For example, the original FC connector had fifteen separate parts that had to be assembled by the technician. Some vendors today offer FC-style connectors with only four separate parts. Such design improvements make the connector easier to use.

Most single-mode connectors use ceramic ferrules. Although a single-mode connector can be used with multimode fibers, a multimode connector should not be used with single-mode fibers. The reason is tolerances. A connector for 125-μm *multimode* fiber must have a ferrule bore large enough to accommodate the largest fiber size of 127 μm. A connector for *single-mode* fiber is often available for the specific tolerance size; that is, you can obtain the connector with a ferrule bore diameter of 125, 126, or 127 μm to fit the fiber exactly. This also means that you may have several connectors on hand to try on the fiber until you get a snug fit. Having a connector with a 126-μm ferrule bore and a fiber with a 127-μm diameter doesn't work.

FC-style Connector

The earliest connector to be based on the 2.5-mm ceramic ferrule, the FC connector was originally devised by Nippon Telephone and Telegraph for telecommunications (Figure 11–9). It has been very popular in Japan, Europe, and the U.S. MCI, for example, used it in its fiber-optic telephone network in the 1980s.

The connector uses a threaded coupling nut, which has the advantage of providing a secure connection even in high-vibration environments and the disadvantage of not permitting quick connection or disconnection. The coupling nut must be rotated several times to thread or unthread the connector.

The connector also offers tunable keying. A keyed connector with a small key results in the ferrule always mating in its adapter bushing the same way. It cannot rotate between one mating and the next; this minimizes any changes due to variations in concentricity or ellipticity in ferrule or fiber (see Figure 11–1). Tunable keying means that the key can be adjusted to the point of lowest loss. For example, insertion loss measurements can be made with the key in several positions. They key is then locked into position.

Although tuning is useful in applications where the lowest possible loss is required, it is not used in most applications because the difference of 0.01 or 0.02 dB offered by tuning is often meaningless.

The connector is available in both single-mode and multimode versions. The earliest version of the FC used a flat ferrule endface, but the newer FC/PC uses a rounded PC endface to permit higher return loss. Some manufacturers offer connectors with an angled endface to achieve the same end.

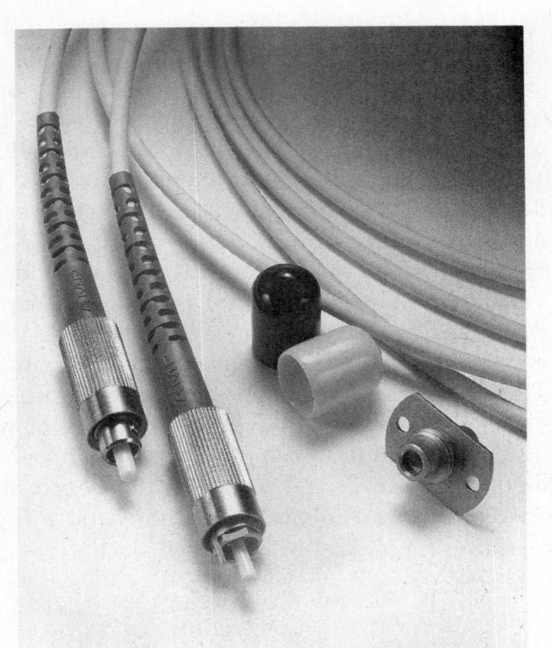

FIGURE 11–9 FC-style and D4-style connectors
(Photo courtesy of AMP Incorporated)

D4-style Connector

This connector is very similar to the FC connector, with its threaded coupling, tunable keying, and PC end finish. The main difference is its 2.0-mm-diameter ferrule. It was originally designed by the Nippon Electric Corporation (Figure 11–9).

ST-style Connector

The ST connector, which was designed by AT&T Bell Laboratories for use in premises wiring of buildings and other applications, uses the same 2.5-mm ceramic ferrule as the PC connector but with a quick-release bayonet coupling (Figure 11–10). Quick-release couplings are preferred in applications where severe vibrations are not expected—such as offices.

With over thirty manufacturers offering ST connectors, it is probably the most popular connector style. It is widely used in local area networks, premises wiring, test equipment, and many other applications. Many applications that specify other connectors allow the ST to be used as an alternative.

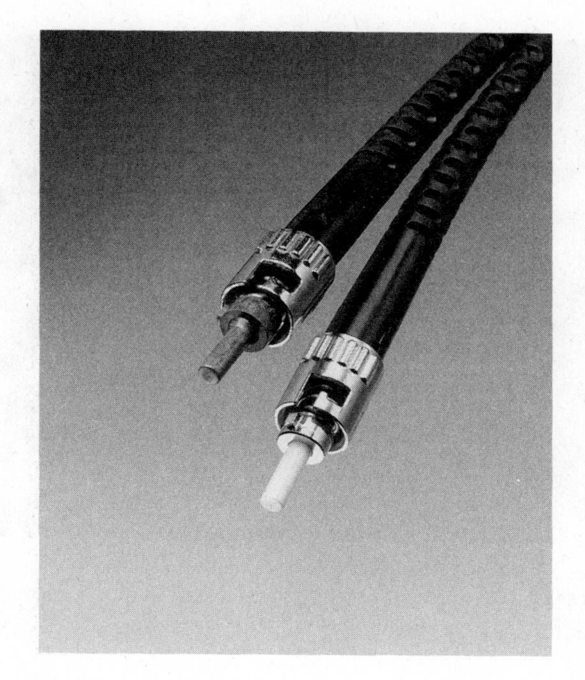

FIGURE 11–10 ST-style connectors
(Photo courtesy of AMP Incorporated)

The quick-release bayonet locking mechanism requires only a quarter turn during mating or unmating, while built-in keying ensures repeatable performance during mating. The key ensures that the fiber is always inserted to the mating bushing with the same orientation: One fiber will not be rotated with respect to another. Often, predictable, consistent loss is more important than lowest loss.

Because of its popularity, the ST connector is offered in several variations: with ceramic, stainless steel, or plastic ferrule; in single-mode and multimode versions; and in epoxyless versions. Insertion loss is around 0.3 dB for a ceramic version to 0.7 for a plastic version. The difference in loss according to material is not caused by the material itself but by the ability to hold precise tolerances during manufacture. In other words, ceramics can be processed to more precise tolerances than plastic.

SC Connectors

The SC connector achieved wide popularity in the early 1990s for both single-mode and multimode applications (Figure 11–11). Originally designed by Nippon Telephone and Telegraph, the connectors use a push-pull engagement for mating.

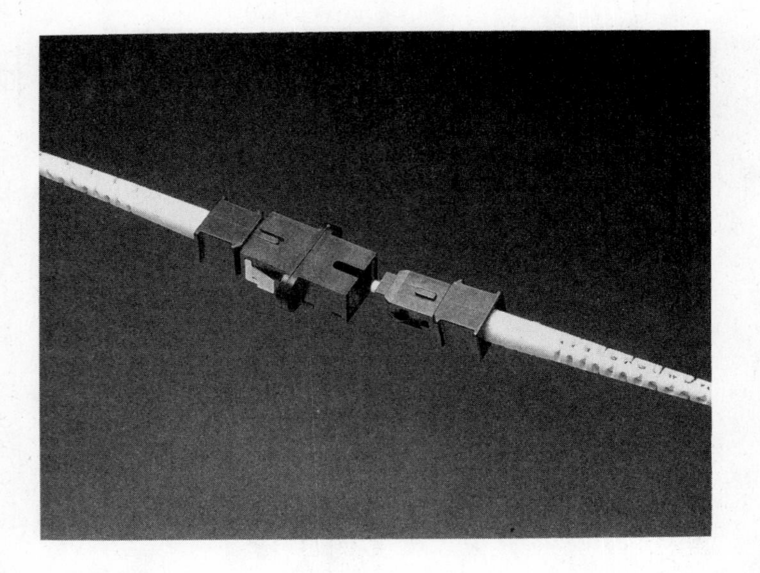

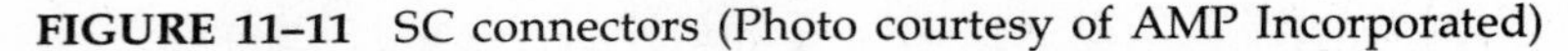

FIGURE 11–11 SC connectors (Photo courtesy of AMP Incorporated)

The SC is designed to be pull-proof. In a pull-proof connector, the ferrule is decoupled from the cable and connector housing. A slight pull on the cable will not pull the ferrule out of optical contact with the interconnection.

The name SC comes from ''subscriber connector,'' which describes its original application in telecommunications. It is replacing the FC and D4 connectors in new telecommunications applications worldwide, and it is a strong competitor to the ST connector in local area networks, premises wiring, and similar applications.

The basic SC connector consists of a plug assembly containing the ferrule. The plugs mate into connector housings. One attractive feature of the SC is the ease with which multifiber connectors are constructed, with multiple-position housings or with clips that hold two or more plugs together. Connectors that require twisting, like the FC or ST, are not adaptable to multifiber applications.

An important application of this multifiber capability is duplex applications, where one fiber carries information in one direction and the other fiber carries information in the other direction. For example, a computer workstation with a fiber-optic connection to the network contains both a transmitter and a receiver. A duplex connector to both permits a single connector to plug into both the transmitter and receiver.

Another connector that provides this capability is the FDDI duplex connector.

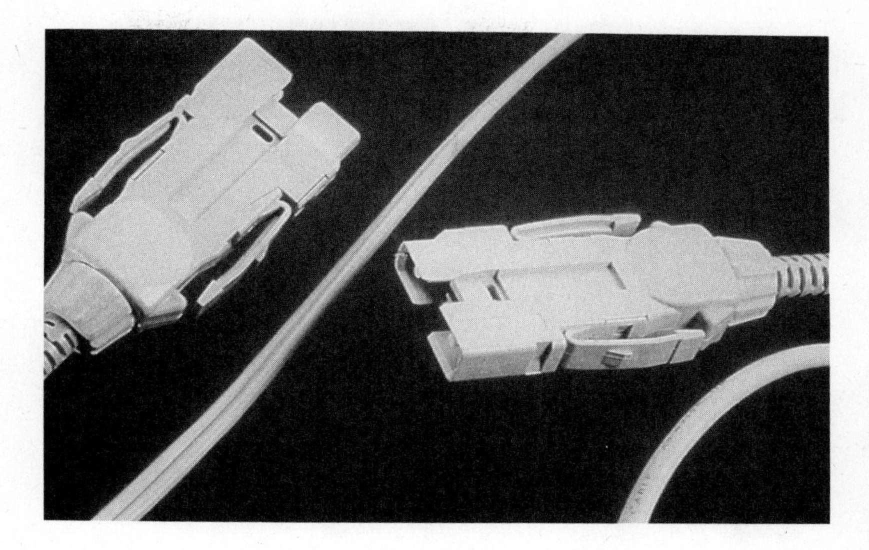

FIGURE 11–12 FDDI MIC connector

FDDI MIC Connector

Designed by ANSI for use in FDDI networks (described in detail in Chapter 15), this connector is a duplex connector using two 2.5-mm ferrules (Figure 11–12). The distinguishing feature is the fixed shroud that protects the ferrules from damage. A floating interface ensures consistent mating without stubbing.

A positive side-latch mechanism and keying capability (per FDDI) make the connector easy to use. Companion coupling adapters allow an FDDI connector to mate with another FDDI connector, with two ST-style connectors, or with a transceiver. The term *MIC* means *medium interface connector,* referring to the connector that serves the interface between electronics and fiber transmission medium.

Because of their features—duplex configuration, low loss, wide availability, and easy use—these connectors are popular beyond FDDI applications.

ESCON Connector

This connector's name derives from its application in IBM's ESCON channel interface mentioned in Chapter 1 and described in Chapter 15 (Figure 11–13). It is similar to the FDDI connector—a duplex connector using 2.5-mm ferrules and a floating interface. The principal difference is its retractable shroud, which pulls

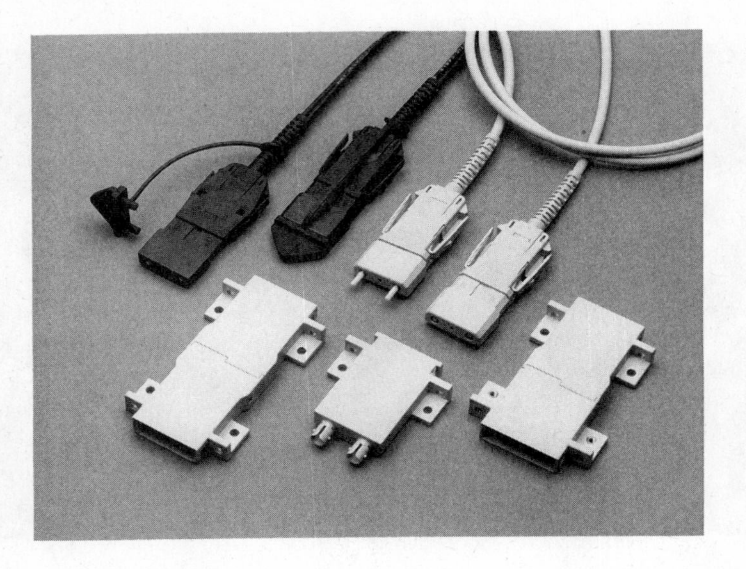

FIGURE 11–13 ESCON connectors (Photo courtesy of AMP Incorporated)

back during engagement of the connector with a transceiver. While this simplifies the design of the transceiver interface, it may not provide as much protection as a fixed shroud.

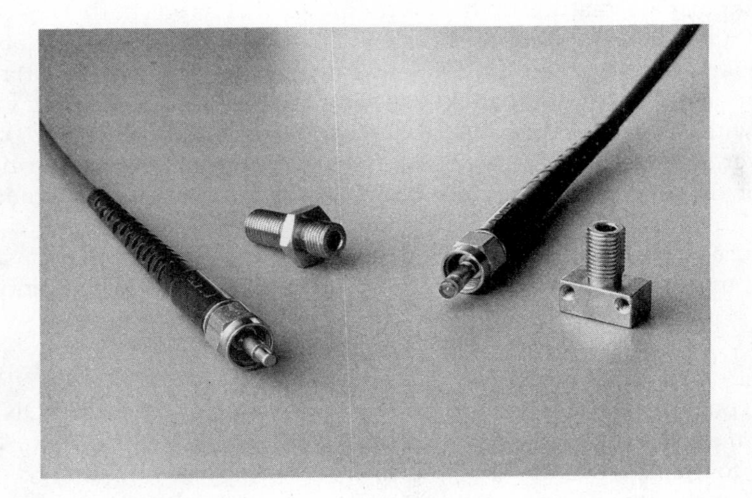

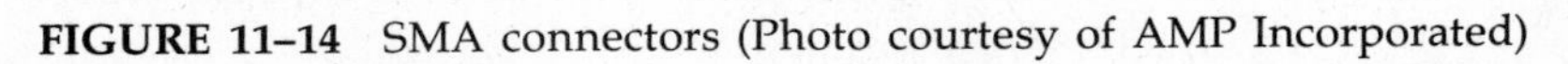

FIGURE 11–14 SMA connectors (Photo courtesy of AMP Incorporated)

SMA Connectors

Connectors using an SMA-style threaded coupling nut are among the most mature and popular connector styles, being standardized by NATO, the U.S. military, and the IEC (Figure 11–14). They were originally designed by the Amphenol Corporation in the late 1970s.

The two basic types of SMA connectors are the 905 style and 906 style. The 905 style uses a straight ferrule. The 906 style has a step-down nose. A plastic alignment bushing fits over the step-down section of mating connectors to help align them. One version of the 905 style uses proprietary resilient tip to align the fibers.

Whereas the original SMA connectors were designed for multimode applications and used a steel ferrule, the connectors are now available with a ceramic ferrule for single-mode applications. Other versions meet the stringent requirements of the MIL-C-83522 standard for military applications.

Plastic-Fiber Connectors

Connectors exclusively for plastic fibers stress *very* low cost and easy application—often with no polishing or epoxy (Figure 11–15). Plastic fiber can be trimmed to an acceptable finish with a hot knife (similar to a cross between an X-acto knife and a soldering iron), obviating the need for polishing. Fiber retention is mechanical, usually by gripping the fiber cladding with barbs or other secure means. This allows very fast application, with much less skill required for acceptable terminations than with glass fibers.

Plastic fiber connectors include both proprietary designs and standard designs such as the digital audio connector described in Electronics Industry of Japan (EIAJ) RC-5720. Connectors such as the ST or SMA are also available for use with plastic fibers. As plastic-fiber technology gains popularity in such applications as digital audio electronics and other consumer applications, there will undoubtedly be greater standardization. In other applications—automotive or security systems, for instance—standardization is less critical.

SPLICES

Fusion Splice

Fusion splicers use an electric arc to weld two fibers together. Fusion splicers offer very sophisticated, computer-controlled alignment of fibers to achieve the lowest losses routinely achieved—as low as 0.05 dB. Because they fuse the fiber, they virtually eliminate return reflections. Their main drawback is the high cost of equipment. Nevertheless, fusion splices remain the choice where the lowest possible losses are required.

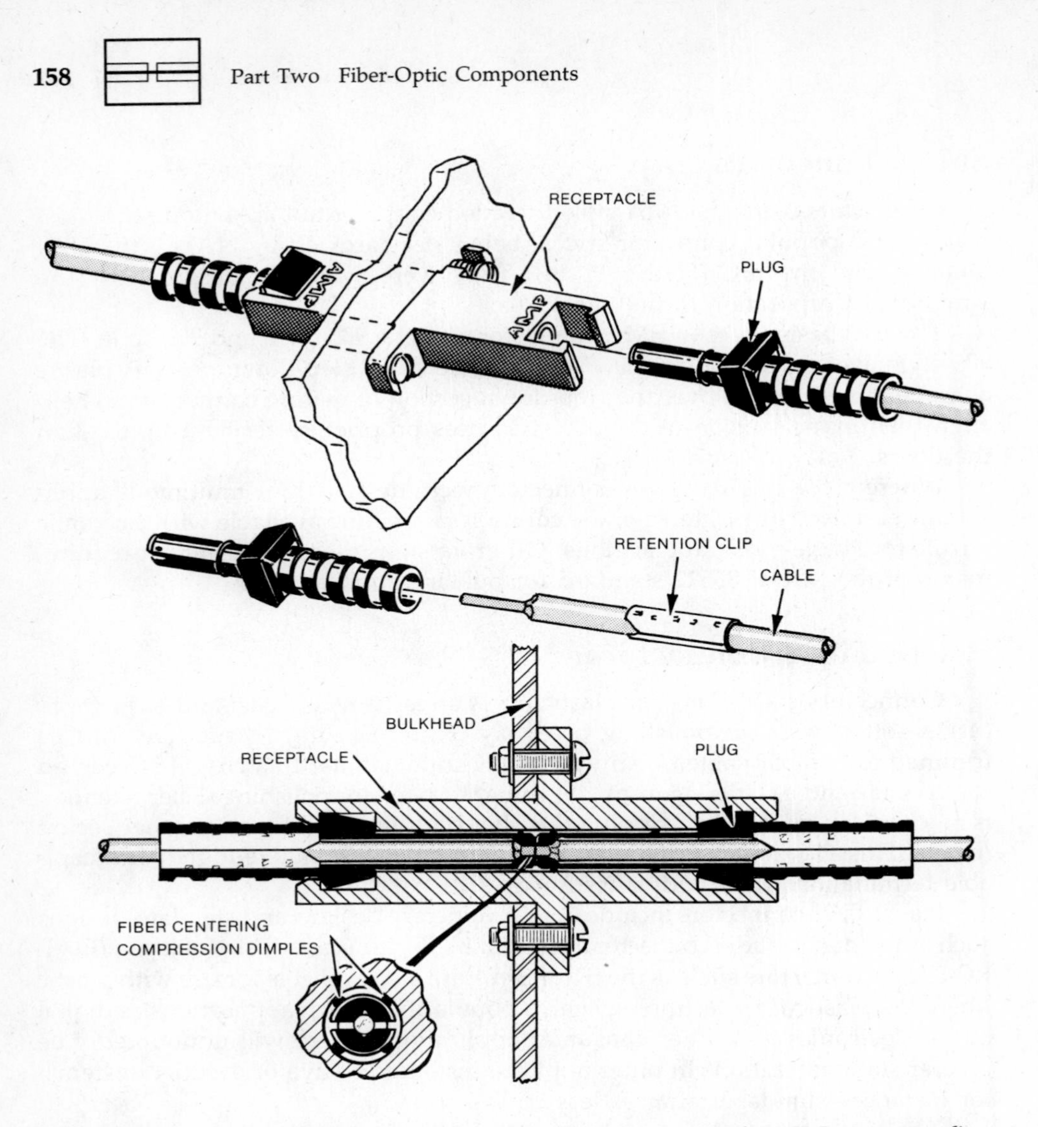

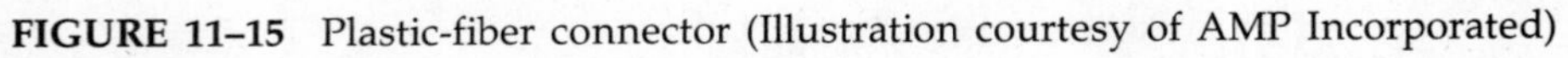
FIGURE 11–15 Plastic-fiber connector (Illustration courtesy of AMP Incorporated)

Mechanical Splices

Several forms of mechanical splices have been devised. These all share common elements: They are easily applied in the field, require simple or no tooling, and offer losses on the order of 0.2 to 0.25 dB. Some splices are reenterable—a telephone industry term meaning that they can be reused.

Figure 11–16 shows a typical splice, the finger splice invented by Siemens of Germany.

While the main movement in connectors is toward offering such industry-standard designs as the ST, SC, or FDDI connector, each splice design remains

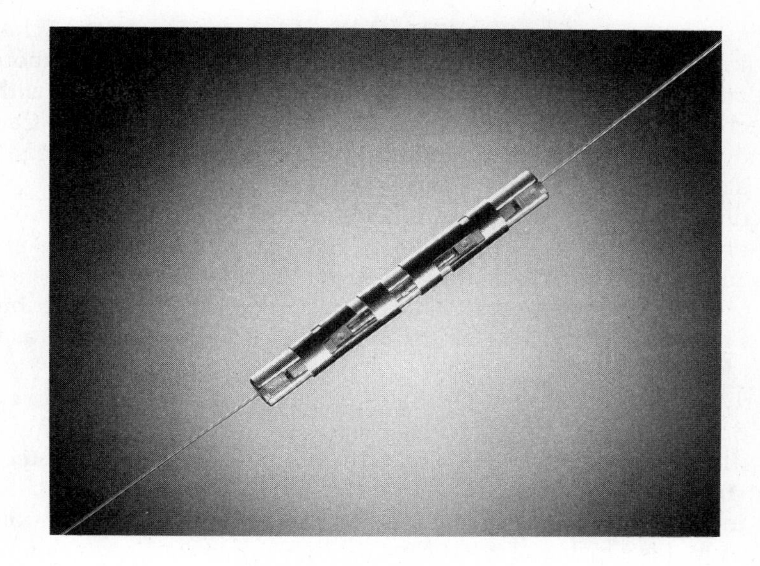

FIGURE 11–16 Finger splice

proprietary to the manufacturer. The reason is simple: They don't have to be compatible because they do not mate with anything.

FIBER PREPARATION

Proper preparation of the fiber end face is critical to any fiber-optic connection. The two main features to be checked for proper preparation in end finish are *perpendicularity* and *end finish*. The end face ideally should be perfectly square to the fiber and practically should be within 1° or 2° of perpendicular. Any divergence beyond 2° increases losses unacceptably. The fiber face should have a smooth, mirrorlike finish free of blemishes, hackles, burrs, and other defects.

The two common methods used to produce correct end finishes are the *scribe-and-break method* (also called the score-and-break method) and the *polish method*. The scribe-and-break method is used mostly with splices, whereas the polish method is more commonly used with connectors.

Whichever method is used, the fibers must first be bared. Jackets, buffer tubes, and other outer layers are removed with wire strippers, cutting pliers, utility knives, and other common tools. Metal strength members are removed with wire cutters, and scissors are used to cut Kevlar strength members.

The plastic buffer coating attached to the cladding can be removed chemically or mechanically. Chemical removal involves soaking the fiber end for about 2 min in a paint stripper or other solvent and then wiping the fiber clean with a soft tissue. Mechanical removal is done with a high-quality wire stripper. Care must be taken not to nick or damage the cladding. The bared fiber can be cleaned with isopropyl alcohol or some other suitable cleaner.

The scribe-and-break method uses a fiber held under slight pressure. A cutting tool with a hard, sharp blade, such as diamond, sapphire, or tungsten carbide, scribes a small nick across the cladding. The blade can be pulled across a stationary fiber, or a fiber can be pulled across a stationary blade. After scribing, the pressure is increased by pulling, which forces the flaw to propagate across the face of the fiber.

Properly done, the cleave produces a perpendicular, mirrorlike finish. Improperly done, the cleave results in hackles and lips, as shown in Figure 11–17, that make the cleave unacceptable. The process must then be repeated.

Scribe-and-break cleaving can be done by hand or by tools that range from relatively inexpensive hand tools to elaborate automated bench tools. Any technique or tool is capable of good cleaves; the trick is consistent finishes time and time again. In general, the less costly approaches require more skill and training for the technicians making the cleave.

Most connectors use polishing to achieve the proper end finish after the fiber has been partially or completely assembled in the connector. Polishing is the final step of assembly. Most connectors use some sort of polishing fixture, which has a large polishing surface to ensure a perpendicular finish. A smaller-faced fixture runs the risk that the assembly will be cocked or tilted during hand polishing.

Polishing is done in two or more steps with repeatedly finer polishing grits, typically down to 1 μm or 0.3 μm. Polishing is done with a figure-8 motion. The connector and fiber face should be cleaned before switching to a finer polishing material.

As with a cleaved fiber, the polished fiber should be inspected under a microscope. Small scratches on the fiber face are usually acceptable, as are small pits on the outside rim of the cladding. Large scratches, pits in the core region, and fractures indicate unacceptable end finishes. Some poor finishes, such as scratches, can be remedied with additional polishing with 1-μm, or finer, film. Fractures and pits usually mean a new connector must be installed. Figure 11–18 shows examples of acceptable and unacceptable polishes.

CONNECTOR ASSEMBLY EXAMPLE

To demonstrate the practical assembly of a connector, we reproduce in this section the assembly instructions for an SC-style connector from AMP Incorporated (Figure 11–19).

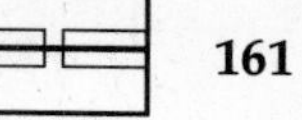

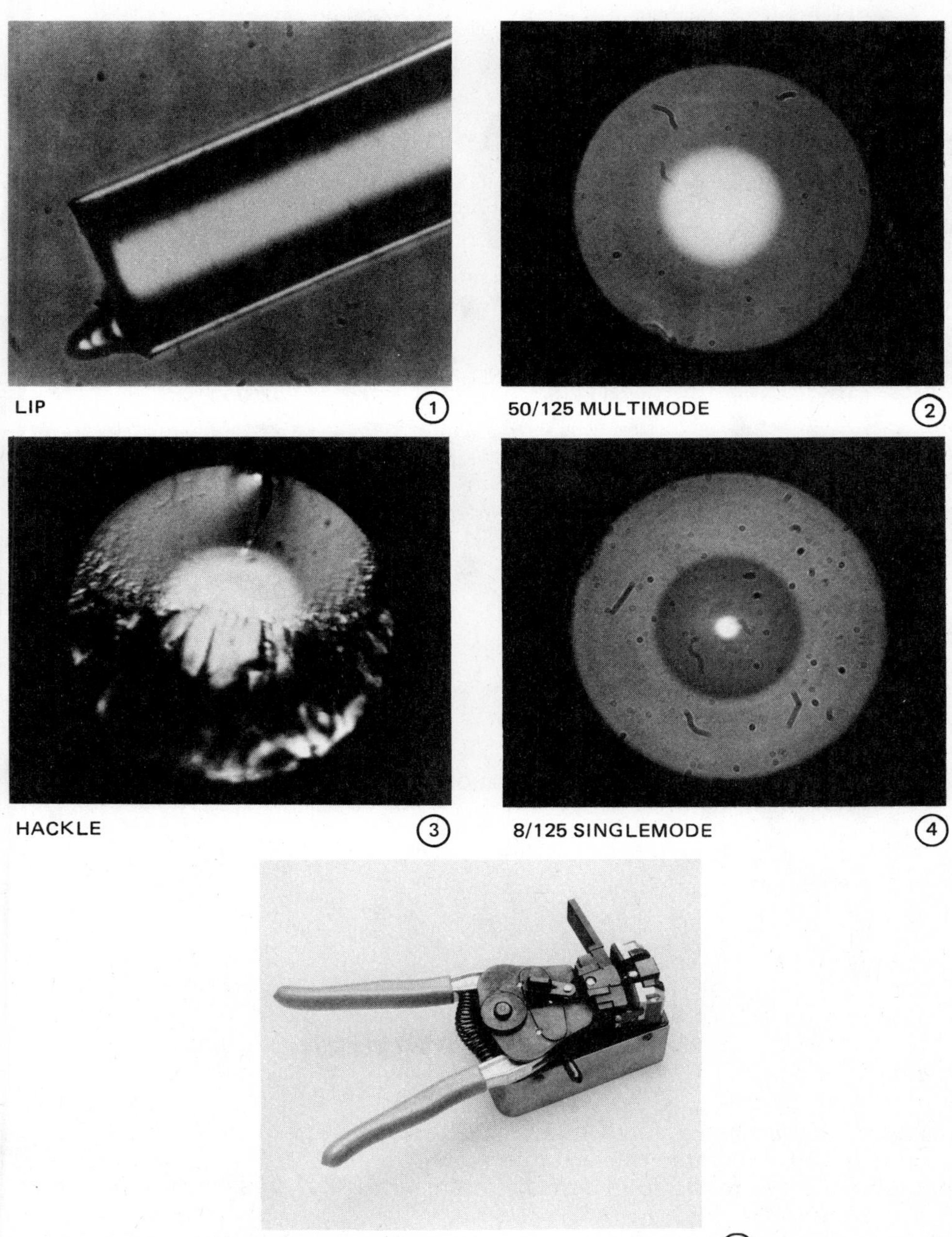

FIGURE 11–17 Good and bad cleaves of an optical glass fiber (Photos courtesy of GTE Fiber Optic Products)

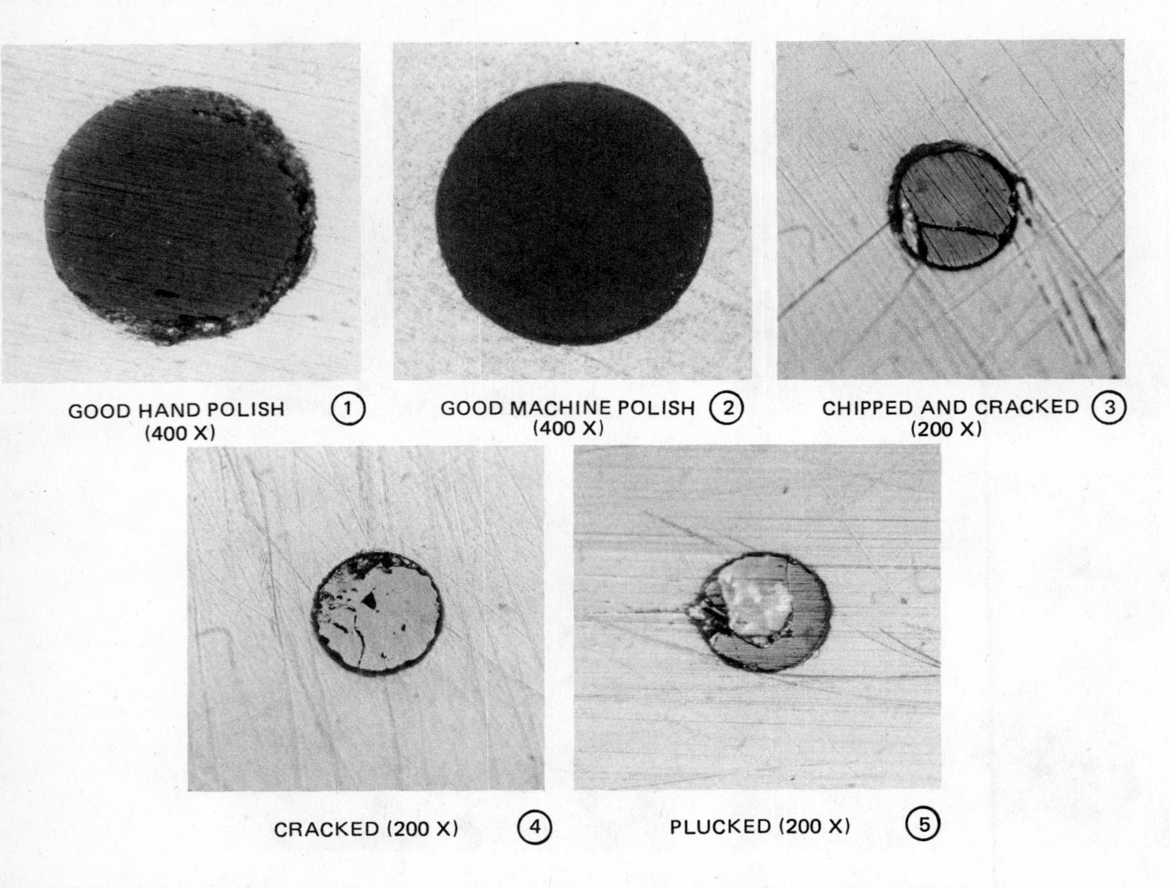

FIGURE 11–18 Good and bad polishes (Photos courtesy of Buehler Ltd., 71 Waukegan Road, Lake Bluff, IL 60044)

SUMMARY

- A splice is used for permanent and semipermanent connections.
- A connector is used for disconnectable connections.
- Loss in an interconnection results from intrinsic factors, extrinsic factors, and system-related factors.
- There are many different designs for connectors and splices. The common ingredient in all designs is precise alignment of fibers.
- Fiber ends can be prepared by the scribe-and-break technique (cleaving) or by polishing.

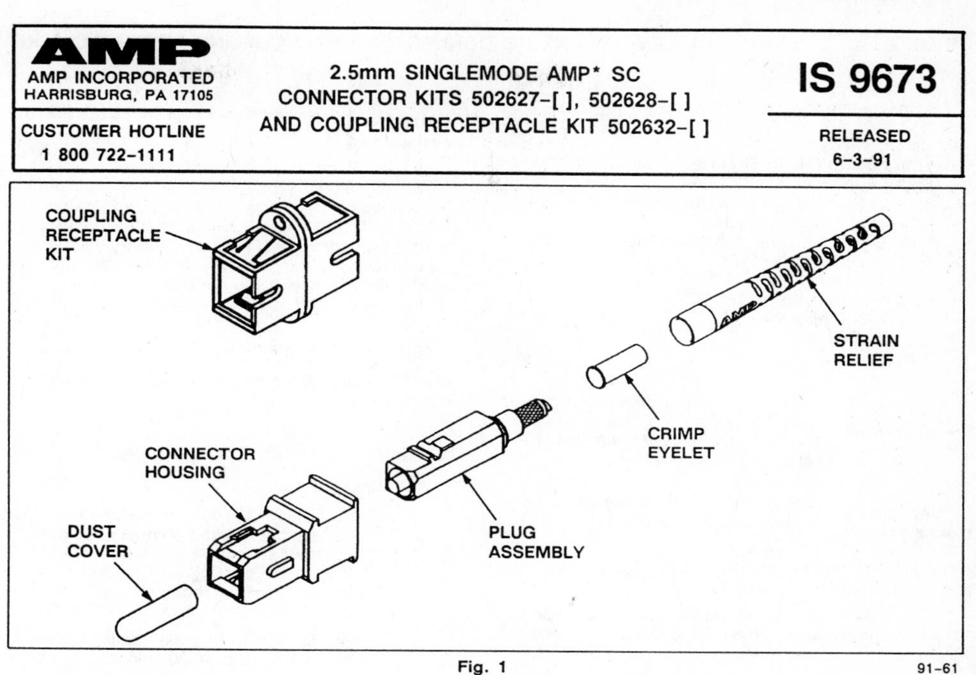

Fig. 1

91-61

1. INTRODUCTION

This instruction sheet (IS) covers the application of Singlemode AMP SC Connector Kits to fiber-optic cable. Kit 502627-[ ] features an alumina ceramic ferrule and kit 502628-[ ] has a zirconia ceramic ferrule. Refer to the respective customer drawings for ferrule hole size and color combinations.

AMP SC Coupling Receptacle Kit 502632-[ ] is also covered. Consult customer drawing 502632 for overall dimensions and color choices.

Read this material thoroughly before starting assembly.

NOTE *Dimensions on this sheet are in millimeters [with inch equivalents in brackets]. Figures and illustrations are for identification only, and are not drawn to scale.*

2. DESCRIPTION (Figure 1)

The connector kits consist of a plug assembly, connector housing, ribbed strain relief, crimp eyelet, and dust cover. These connectors are used in conjunction with 125-μm Singlemode fiber-optic cable having a jacket diameter of 3mm [.12 inch].

The coupling receptacle is used to mate two AMP SC Connectors. A receptacle assembly and panel clip make up the receptacle. The receptacle can be used in free-hanging applications, or it can be mounted to a panel. See Figures 11 and 12 for mounting information.

3. ASSEMBLY PROCEDURE

A. Required Tools and Materials

The following tools and materials are required for applying the connectors to optical fibers:

Tools:

— Cable Stripper 501198-1 (IS 9394)
— Crimping Tool 58190-6 with Die Assembly 58289-1 (IS 9047); or Hand Tool 220190-1 (IS 2901) with Die Assembly 58299-1 (IS 6969)
— Epoxy Mixer 501202-1
— Fiber Stripper 501013-2 (IS 9485)
— Microscope 501196-5 (IS 9111) or Magnifier 501095-1
— Polishing Bushing 502631-1
— Polishing Plate 501197-2
— Fiber Protector 502656-1
— Scissors 501014-1
— Scribe Tool 502684-1 (IS 9697)
— Cable Preparation Template Kit 501818-1

*Trademark of AMP Incorporated

1 OF 8

FIGURE 11–19 Connector assembly example

IS 9673 2.5mm SINGLEMODE AMP SC CONNECTOR KITS/COUPLING RECEPTACLE KIT

Consumable Items:

- Cotton swabs
- Epoxy 501195-4, or 502418-1 (Fast Cure)
- Epoxy Applicator Kit 501473-3
- Isopropyl Alcohol (Alcohol pads are recommended)
- Lint-Free Tissues or Cloths
- Polishing Pad 501858-1 (Green Pad)
- 5-µm Polishing Film 228433-8
- .3-µm Polishing Film 228433-5
- Final Polishing Film 502748-1 (Single Sheet), 502748-2 (Five Pack)

B. Preparing Fibers

DANGER *Be very careful to dispose of fiber ends properly. The fibers create slivers that can easily puncture the skin and cause irritation.*

DANGER *Always wear safety glasses when working with optical fibers.*

1. Cut the cable 25mm [1 in.] longer than the required finished length.

2. Slide strain relief onto the cable; then slide on the crimp eyelet with the flange toward the end of the cable. See Figure 1.

3. Strip the cable to the dimensions shown in Figure 2 using cable stripper 501198-1, fiber stripper 501013-2 and scissors 501014-1. Use OPTIMATE 2.0, 2.5mm Threaded and AMP SC Connector template in cable preparation template kit 501818-1 for recommended strip dimensions and tolerances.

4. Clean the fiber thoroughly using an alcohol pad.

CAUTION *Never clean a fiber with a dry tissue.*

5. Obtain a "slip fit" between the fiber and the connector by fitting various sized connectors (125, 126, 127-µm) to the fiber. Start with the largest connector (127-µm); then try the 126-µm connector, etc., until you find the size that fits the fiber snugly. If the connector does not fit the fiber, do not force it. Use the previous, larger connector.

6. Repeat step 5 for each end to be terminated.

7. Evenly "fan out" the strength members from the buffer.

C. Selecting and Preparing Epoxy

1. Select epoxy:

   Epoxy 501195-4 is easy to work with because it comes in packs with pre-measured components. It will cure in 24 hours at 25°C [72°F] or 2 hours at 65°C [150°F]. Use of Epoxy Mixer 501202-1 is recommended for thorough mixing of the components.

   Epoxy 502418-1 is bulk packaged in two tubes. It will cure in 30 minutes at 100°C [212°F].

2. Preparing epoxy 501195-4:

   a. Remove the separating clip from epoxy package and mix epoxy thoroughly for 20-30 seconds using epoxy mixer 501202-1.

   b. Install the needle tip on the epoxy applicator. Make sure it is secure. Remove the plunger.

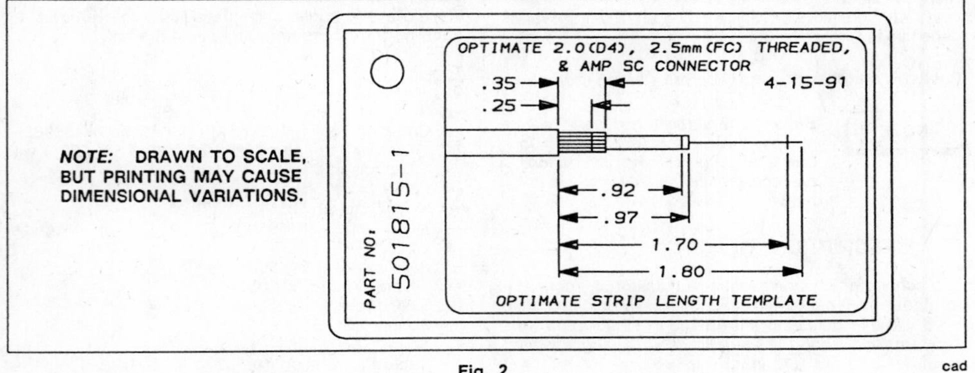

Fig. 2

cad

2 OF 8

FIGURE 11–19 (continued)

2.5mm SINGLEMODE AMP SC CONNECTOR KITS/COUPLING RECEPTACLE KIT IS 9673

c. Cut the epoxy packet diagonally at one corner and squeeze the epoxy into the back of the applicator. Replace the plunger. Hold the applicator vertically (needle upwards), and slowly push on the plunger until the entrapped air escapes and a bead of epoxy appears at the tip.

3. Preparing epoxy 502418-1 (fast cure):

a. Squeeze equal length lines of the two components onto a clean disposable surface.

b. Mix thoroughly with a wooden stick.

c. Install the needle tip on the epoxy applicator. Make sure it is secure. Remove the plunger.

d. Load the applicator from the back. Replace the plunger. Hold the applicator vertically (needle upwards), and slowly push on the plunger until the entrapped air escapes and a bead of epoxy appears at the tip.

D. Terminating Fibers

1. Hold the connector vertically with the ceramic facing downward. Clean the applicator tip and insert it into the tubing until it bottoms. See Figure 3A.

NOTE *Do not get epoxy on the outside of the tubing.*

2. Slowly inject epoxy through the applicator until the epoxy first appears at the ceramic tip of the connector. Withdraw the applicator approximately 1.5mm [1/16 in.] from its bottomed position and inject a very small amount of epoxy.

3. Withdraw the applicator quickly, without injecting additional epoxy into the connector.

If too much epoxy is injected into the connector, it will not function properly.

4. Push the ceramic tip of the connector down on a soft surface so that the tubing protrudes an additional 3mm [1/8 in.] out of the back of the connector. See Figure 3B.

5. Cut the tubing flush with the back of the connector. See Figure 3B.

CAUTION *Additional epoxy must NOT spill out of the tube or get onto any part of the connector. The tube must be able to move inside the connector after the epoxy is hard.*

6. Carefully insert the cable into the connector until the strength members and the jacket bottom against the connector end. As you insert the cable, the fiber should appear at the ceramic tip.

Do not apply any epoxy to the knurled area.

7. Slide crimp eyelet against the connector shoulder, trapping strength members against the knurl. See Figure 3C.

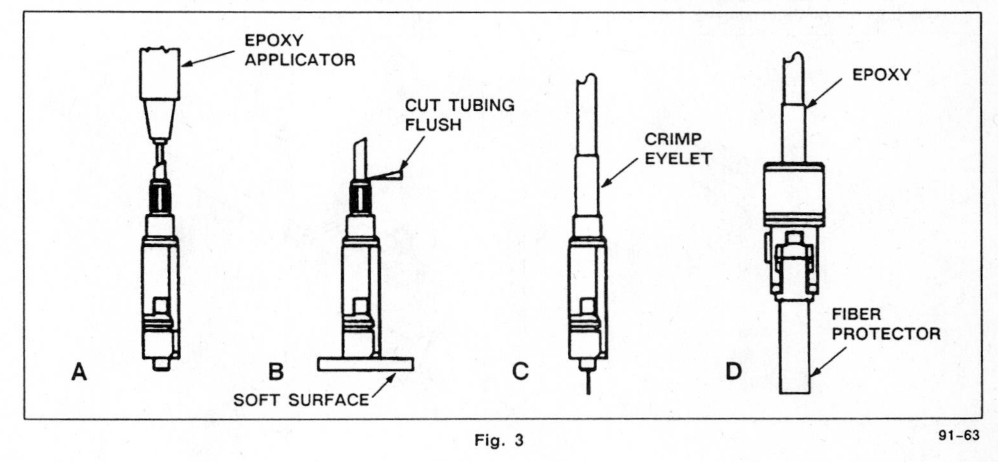

Fig. 3

91-63

3 OF 8

FIGURE 11–19 (continued)

IS 9673 **2.5mm SINGLEMODE AMP SC CONNECTOR KITS/COUPLING RECEPTACLE KIT**

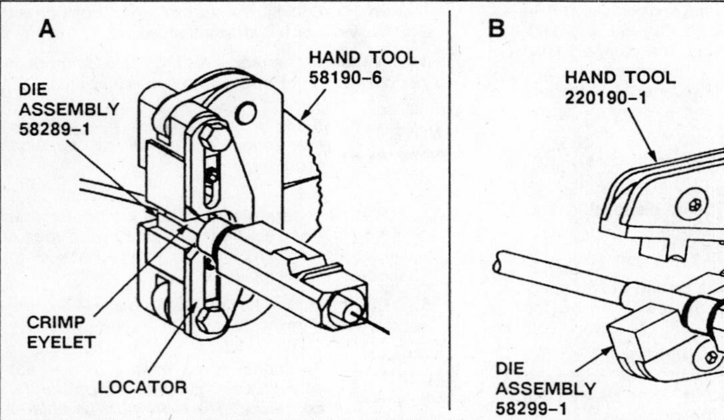

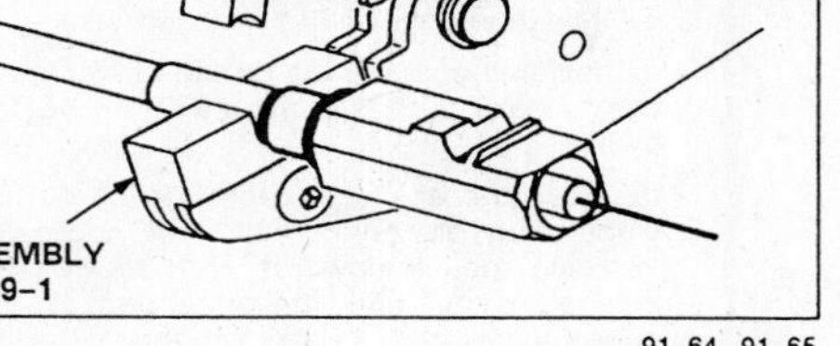

Fig. 4 91-64, 91-65

E. Crimping

— **Crimping with Hand Tool 58190-6**

NOTE *See IS 9047, packaged with Hand Tool 58190-6, and Die Assembly 58289-1, for detailed instructions.*

1. Install Die Assembly 58289-1 in Hand Tool 58190-6.

2. Squeeze the handles on the hand crimping tool until the ratchet releases. Open the tool fully.

3. Place the connector in the dies so that the crimp eyelet rests in the die against the locator. See Figure 4A.

4. Squeeze the crimping tool handles shut to crimp the eyelet.

5. Slide the connector housing over the connector body until it clips into place. Orient the lug on the connector housing with the chamfers on the connector assembly as shown in Figure 5.

6. Dip the end of a small piece of scrap fiber (without buffer) into the epoxy pool and apply a small drop at the fiber-to-ceramic ferrule interface. The bead should be no larger than 1mm [.04 in.]. See Figure 6.

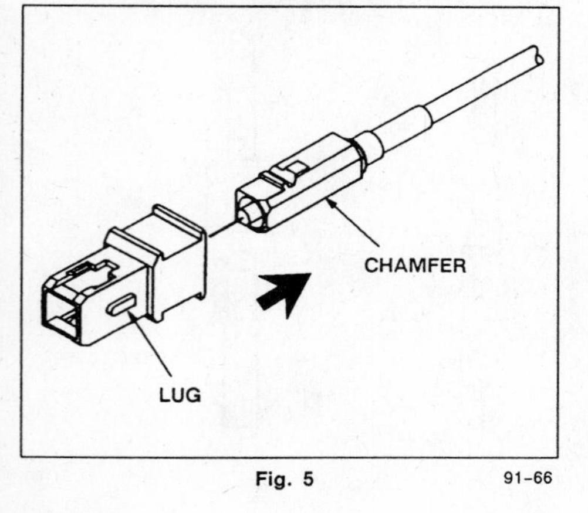

Fig. 5 91-66

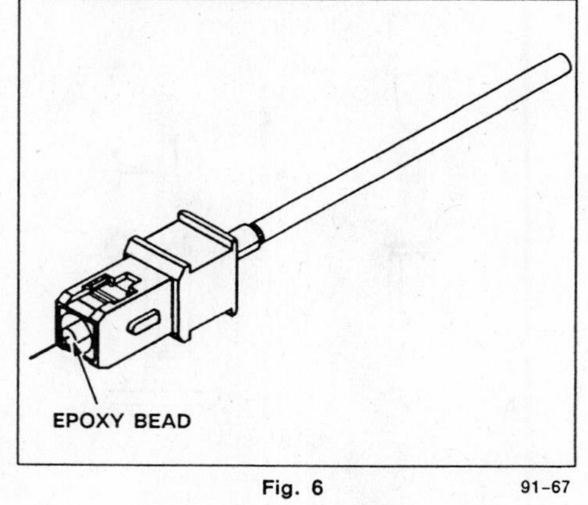

Fig. 6 91-67

FIGURE 11–19 (continued)

2.5mm SINGLEMODE AMP SC CONNECTOR KITS/COUPLING RECEPTACLE KIT IS 9673

7. Install a fiber protector over the fiber/connector at this time.

8. Apply a thick layer of epoxy to the crimp eyelet/cable jacket interface. See Figure 3D.

9. Slide the strain-relief over the crimp eyelet and push the strain-relief towards the connector until it bottoms.

10. Hang the connector vertically with the tip down and cure at either:

- 65°C [150°F] for 2 hours or 25°C [72°F] for 24 hours using epoxy 501195-4 or
- 100°C [212°F] for 30 minutes when using epoxy 502418-1.

11. Remove any epoxy from the crimping dies using an alcohol-dampened tissue.

— Crimping with Hand Tool 220190-1

NOTE *See IS 2901, packaged with Hand Tool 220190-1, and IS 6969 packaged with Die Assembly 58299-1, for detailed instructions.*

1. Install Die Assembly 58299-1 in Hand Tool 220190-1.

2. Squeeze the handles on the hand crimping tool until the ratchet releases. Open the tool fully.

3. Place the connector in the dies so that the flange on the crimp eyelet is against the side of the dies. See Figure 4B.

4. Squeeze the crimping tool handles shut to crimp the eyelet.

5. Slide the connector housing over the connector body until it clips into place. Orient the lug on the connector housing with the chamfers on the connector assembly as shown in Figure 5.

6. Dip the end of a small piece of scrap fiber (without buffer) into the epoxy pool and apply a small drop at the fiber to ceramic ferrule interface. The bead should be no larger than 1mm [.04 in.]. See Figure 6.

7. Install a fiber protector over the fiber/connector at this time.

8. Apply a thick layer of epoxy to the crimp eyelet/cable jacket interface. See Figure 3D.

9. Slide the strain-relief over the crimp eyelet and push the strain-relief towards the connector until it bottoms.

10. Hang the connector vertically with the tip down and cure at either:

- 65°C [150°F] for 2 hours or 25°C [72°F] for 24 hours using epoxy 501195-4 or
- 100°C [212°F] for 30 minutes when using epoxy 502418-1.

11. Remove any epoxy from the crimping dies using an alcohol-dampened tissue.

F. Polishing the Fiber

Polish by hand using polishing bushing 502631-1.

1. Remove the fiber protector from the connector.

2. Firmly support the connector assembly.

DANGER *SAFELY DISPOSE OF THE EXCESS FIBER. A handy method is to put a small piece of masking tape on the fiber before scribing it. The fiber is then easily retrieved and disposed of.*

3. Use scribe tool 502684-1 to scribe the fiber. Draw the beveled edge of the tool across the fiber as shown in Figure 7. After scoring the fiber, pull fiber straight away from connector to finish the cleaving process.

Do not allow the scribe tool to make contact with the epoxy. This may damage the sapphire tip.

4. Gently push on the ceramic tip to insure tip movement. If the tip does not move in an axial direction, too much epoxy was used and the cable must be reterminated.

5. Install the connector into polishing bushing 502631-1. See Figure 8.

6. Place the polishing pad 501858-1 on the polishing plate 501197-2. The polishing pad and plate are used throughout the entire polishing procedure.

7. Place the 5-μm polishing film on the polishing pad. Beginning with light pressure, polish the endface of the ceramic on the polishing film using a figure-8 pattern. See Figure 9.

8. Polish on the 5-μm film until the epoxy turns light blue.

9. Clean the ferrule tip and polishing bushing with an alcohol-soaked swab or alcohol pad. Dry the endface.

10. Replace the 5-μm film with .3-μm polishing film, and polish the endface using a figure-8 pattern until all the epoxy is removed.

FIGURE 11–19 (continued)

IS 9673 2.5mm SINGLEMODE AMP SC CONNECTOR KITS/COUPLING RECEPTACLE KIT

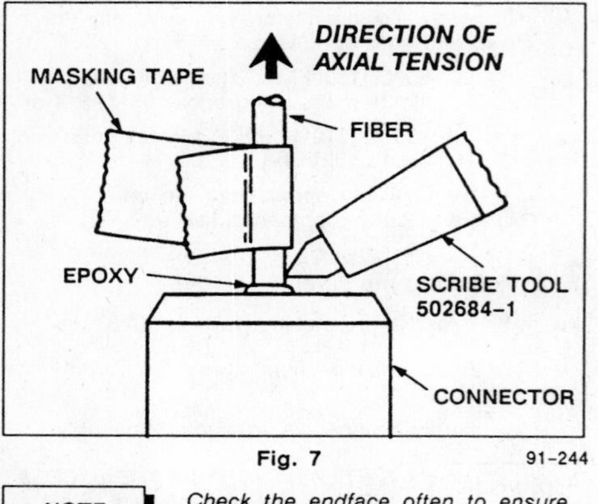

Fig. 7 91-244

NOTE *Check the endface often to ensure polishing ceases as soon as all the epoxy is removed.*

11. Clean the ferrule tip and polishing bushing with an alcohol-soaked swab or alcohol pad. Dry the endface.

12. Replace the .3-µm film with the final polishing film 502748-1. Place a few drops of water on an unused area of film.

13. Polish the connector in the wet area, in circles approximately 19- 20mm [3/4 in.] diameter for 25 seconds.

NOTE *One sheet of final polishing film should be enough for 10-20 connectors.*

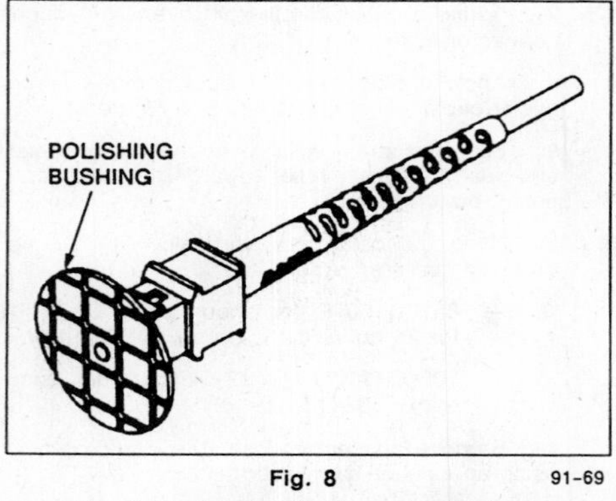

Fig. 8 91-69

14. Clean the ferrule tip and polishing bushing with an alcohol-soaked swab or alcohol pad. Dry the endface.

G. Inspecting the Fiber

DANGER *Never inspect or look into the end of a fiber when optical power is applied to the fiber. The infrared light used, although it cannot be seen, can cause injury to the eyes.*

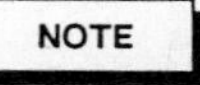

NOTE *See IS 9111 for information on the use of AMP Microscope 501196-5.*

— Be sure all epoxy is removed from the tip.

— Dirt may be mistaken for small pits; if dirt is evident, clean with an alcohol-soaked swab, and then dry.

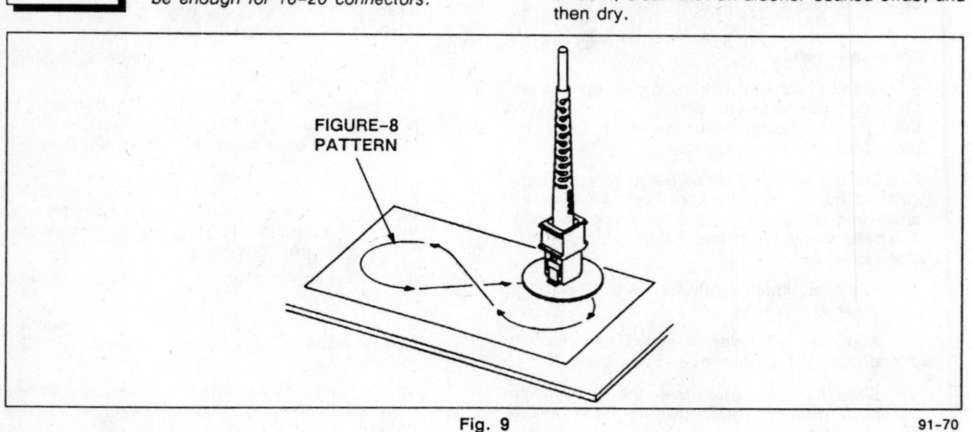

Fig. 9 91-70

FIGURE 11–19 (continued)

2.5mm SINGLEMODE AMP SC CONNECTOR KITS/COUPLING RECEPTACLE KIT **IS 9673**

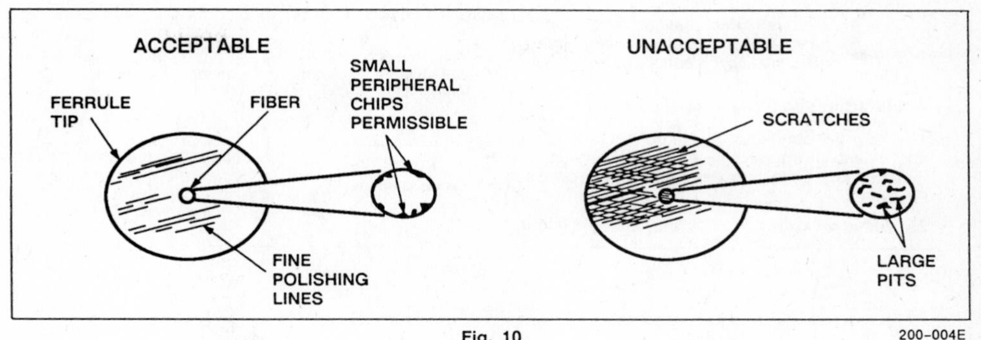

Fig. 10 200-004E

— Large scratches on the fiber endface indicate too much pressure was used when polishing on the 5-μm film or the connector was over-polished on the .3-μm film.

— Fine polishing scratches are acceptable. See Figure 10.

— Small chips at the outer rim of the fiber are acceptable. Large chips in the center of the fiber render the polish unacceptable, and the fiber must be reterminated.

H. Using The Coupling Receptacle

The coupling receptacle can be used free-hanging, or can be mounted to a panel. Two methods of panel mounting are:

First Method

1. Using the cutout dimensions in Figure 11, prepare the panel including the two mounting screw holes.

2. Insert the receptacle through the opening and secure it with two self-tapping screws or use two No. 2 screws and nuts.

Second Method

1. Using the cutout dimensions in Figure 11, prepare the panel, but disregard the mounting screw holes.

2. Insert the receptacle through the opening until the retention clip snaps into place. See Figure 12.

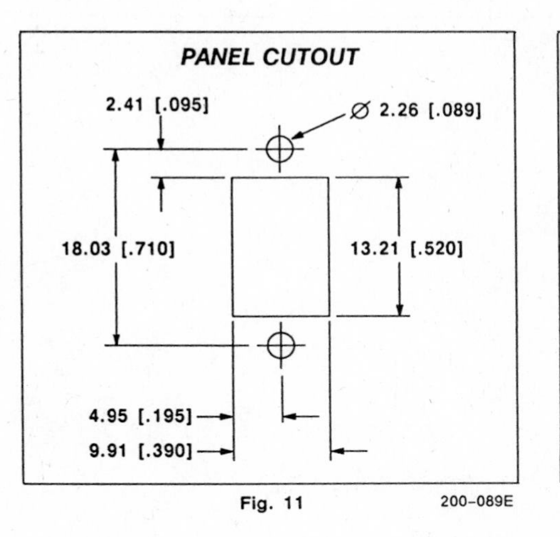

Fig. 11 200-089E

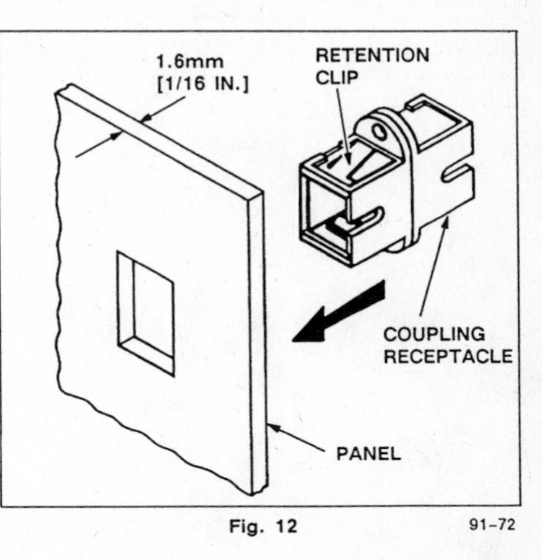

Fig. 12 91-72

7 OF 8

FIGURE 11–19 (continued)

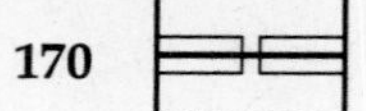

IS 9673 **2.5mm SINGLEMODE AMP SC CONNECTOR KITS/COUPLING RECEPTACLE KIT**

NOTE *After repeated matings, the coupling receptacle should be cleaned (flushed) with isopropyl alcohol and dried.*

I. Connecting To Coupling Receptacle

1. Hold the connector by the connector housing and orient the assembly in the receptacle so that the key goes into the receptacle slot. See Figure 13.
2. Insert until an audible locking "snap" is heard.

J. Disconnect From Coupling Receptacle

Hold the connector by the connector housing and pull straight back until the connector disengages. See Figure 13.

You must use the connector housing to couple/uncouple the connector. Using the strain-relief for this purpose will damage the connector.

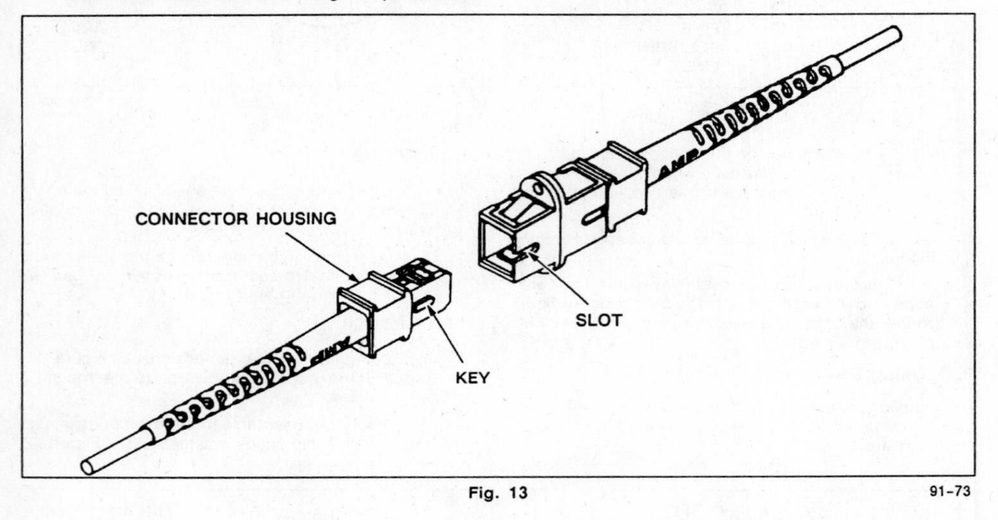

Fig. 13

91-73

FIGURE 11–19 (continued)

? REVIEW QUESTIONS

1. What is the purpose of a splice or connector?
2. What role does the alignment mechanism play in a fiber-optic connector or splice?
3. Name three sources of intrinsic loss in an interconnection.
4. Name three sources of extrinsic loss in an interconnection.
5. Describe how modal patterns affect loss in a fiber-to-fiber interconnection. Will the loss probably be lower if the transmitting fiber is fully filled or at EMD? Why?
6. Assume an application requiring five 10-km reels of fiber to be joined in a 50-km link. Would connectors or splices be preferred? Why?
7. Would a connector or splice be preferred to connect a fiber-optic cable to a personal computer? Why?
8. Calculate the total mismatch loss for a transmitting fiber with a 62.5/125-μm diameter and a 0.29 NA and a receiving fiber with an 85/125-μm diameter and a 0.26 NA. Ignore other sources of loss for this example.
9. What is the insertion loss test meant to measure? What does the test attempt to eliminate as a loss factor?
10. Name the two critical results a good cleave must achieve.

CHAPTER 12

Couplers

Thus far, we have viewed the fiber-optic link as a point-to-point system—one transmitter linked to one receiver over an optical fiber. Even a duplex link is point to point—one transceiver communicates with another transceiver over a fiber-optic pair. In many applications, however, it is desirable or necessary to divide light from one fiber into several fibers or, conversely, to couple light from several fibers into one fiber. In short, we wish to distribute the light. A *coupler* is a device that performs such distribution. Figure 12–1 shows the idea of a coupler: It divides or combines light. The figure also shows examples of actual devices, including couplers, WDM couplers (discussed later in this chapter), and transmitters and receivers.

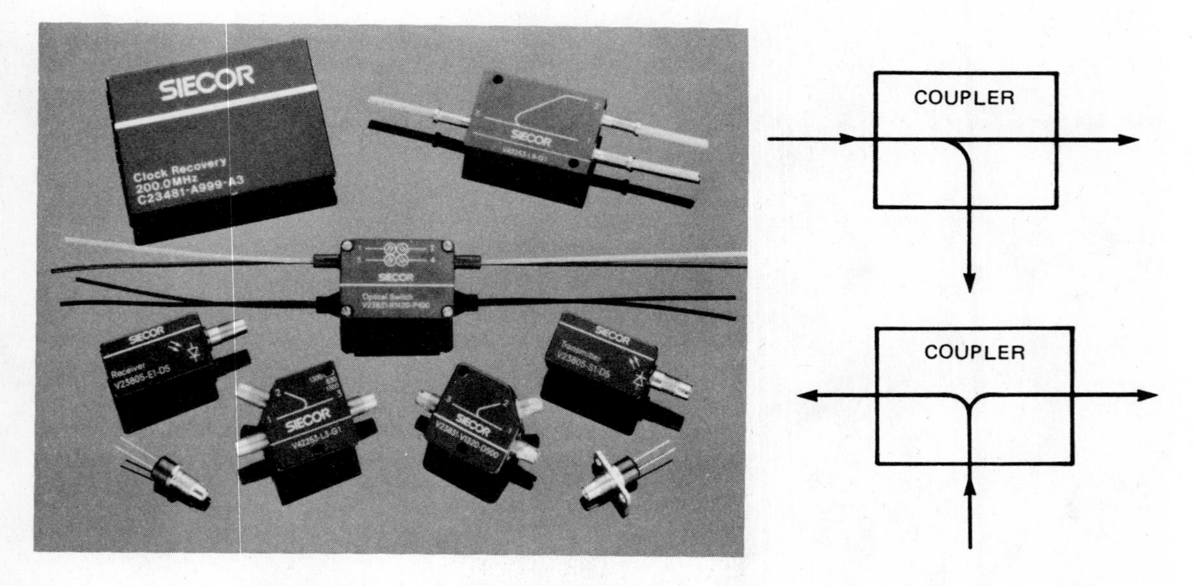

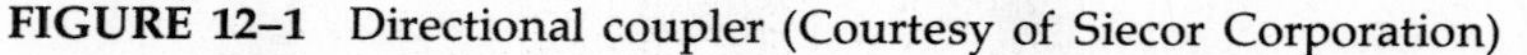
FIGURE 12–1 Directional coupler (Courtesy of Siecor Corporation)

This chapter describers couplers and their uses. Important application areas for couplers are in networks, especially local area networks, and in wavelength-division multiplexing (WDM).

COUPLER BASICS

A *coupler* is a multiport device. A *port* is an input or output point for light. There are several types of loss associated with a coupler. Figure 12–2 shows a four-port directional coupler that we will use to define ideas important to your understanding of couplers. Arrows indicate the possible directions of flow for optical power through the coupler. Light injected into port 1 will exit through ports 2 and 3. Ideally, no light will appear at port 4. Similarly, light injected into port 4 will also appear at ports 2 and 3 but not at port 1.

The coupler is passive and bidirectional. Ports 1 and 4 can serve as input ports, ports 2 and 3 as output ports. Reversing the direction of power flow allows ports 2 and 3 to serve as input ports and port 1 and 4 to serve as output ports.

For the following discussion of loss, we assume port 1 is the *input* port and ports 2 and 3 are the *output* ports. Furthermore, the power at port 2 is always equal to or greater than the power at port 3. Therefore, we will term port 2 as the *throughput* port. Port 3 is the *tap* port. These terms are used to suggest that a path containing the greater part of the power is the throughput path, whereas the path containing the lesser part is the tapped path.

Throughput loss is the ratio of output power at port 2 to input power at port 1:

$$\text{loss}_{\text{THP}} = 10 \log_{10} \left(\frac{P_2}{P_1} \right)$$

Tap loss is the ratio of the output power at port 3 to the input power at port 1:

$$\text{loss}_{\text{TAP}} = 10 \log_{10} \left(\frac{P_3}{P_1} \right)$$

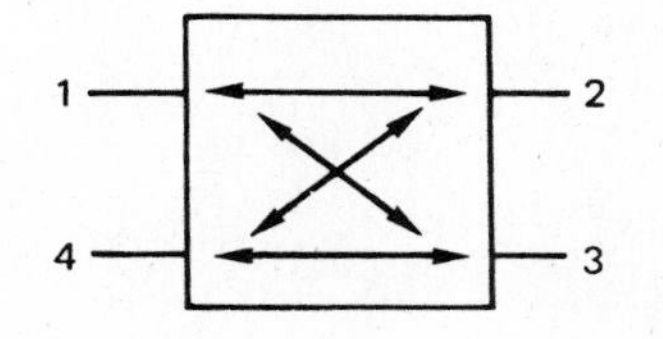

FIGURE 12–2 Four-port directional coupler

Directionality is the ratio between unwanted power at port 4 and the input power at port 1:

$$\text{loss}_D = 10 \log_{10} \left(\frac{P_4}{P_1} \right)$$

Ideally, no power appears at port 4, so $\text{loss}_D = 0$. Practically, some power does appear through such mechanisms as leakage or reflections. Directionality is sometimes called *isolation.* Directionality or isolation in a good tap is 40 dB or greater; only a very small amount of light appears at port 4.

Excess loss is the ratio between output power at ports 2 and 3 to the input power at port 1:

$$\text{Loss}_E = 10 \log_{10} \left(\frac{P_2 + P_3}{P_1} \right)$$

Excess losses are losses that occur because the coupler is not a perfect device. Losses occur within fibers internal to the coupler from scattering, absorption, reflections, misalignment, and poor isolation. In a perfect coupler, the sum of the output power equals the input power ($P_2 + P_3 = P_1$). In a real coupler, the sum of the output power is always something less than the input power because of excess loss ($P_2 + P_3 < P_1$).

Excess loss does not include losses from connectors attaching fibers to the ports. We looked at these losses and their causes in the last chapter. Furthermore, since most couplers contain an optical fiber at each port, additional loss can occur because of diameter and NA mismatches between the coupler port and the attached fiber.

The input power must obviously be divided between the two output ports. The *coupler splitting ratio* is simply the ratio between the throughput port and the tap port: P_2/P_3. Typical ratios are 1:1, 2:1, 3:1, 6:1, and 10:1. From the splitting ratio, the relation of the throughput loss and tap loss is constant. Table 12–1 shows the losses for perfect taps of given splitting ratios.

Splitting Ratio	Throughput Loss (dB)	Tap Loss (dB)
1:1	3	3
2:1	1.8	4.8
3:1	1.25	6
6:1	0.66	8.5
9:1	0.46	10
10:1	0.41	10.4

TABLE 12–1 Losses for ideal four-port directional couplers

In a real coupler, the losses at the output ports are the sums of the individual loss and the excess loss. If $\text{Loss}_{\text{THP}'}$ and $\text{Loss}_{\text{TAP}'}$ represent throughput and tap loss in a real coupler, actual losses become:

$$\text{Loss}_{\text{THP}'} = \text{Loss}_{\text{THP}'} + \text{Loss}_{\text{E}}$$
$$\text{Loss}_{\text{TAP}'} = \text{Loss}_{\text{TAP}'} + \text{Loss}_{\text{E}}$$

Assume a directional coupler with a 3:1 splitting ratio and an excess loss of 1 dB. Actual losses become 2.25 dB for throughput power and 8 dB for tap power. If 100 μW of optical power are input to port 1, how much output power appears at each port? Since loss is the ratio of power at the two ports, we know that the throughput power at port 2 equals

$$2.25 \text{ dB} = 10 \log_{10} \left(\frac{P_2}{100 \ \mu\text{W}} \right)$$

The equation, notice, accounts for excess loss. The 2.25 dB includes 1.25 dB throughput loss plus 1 dB for excess loss. We rearrange and solve the equation to find the power at port 2:

$$P_2 = \left[\log_{10}{}^{-1} \left(\frac{-2.25 \text{ dB}}{10} \right) \right] 100 \ \mu\text{W}$$
$$= 60 \ \mu\text{W}$$

(Notice that since loss is a negative quantity, we make the 2.25 dB negative in the equation.) Tap power at port 3 is

$$P_3 = \left[\log_{10}{}^{-1} \left(\frac{-7 \text{ dB}}{10} \right) \right] 100 \ \mu\text{W}$$
$$= 20 \ \mu\text{W}$$

A directional coupler is symmetrical, so that the losses remain the same regardless of which ports serve as the input port, throughput port, tap port, and isolated port.

TEE COUPLER

A *tee coupler* is a three-port device. Figure 12–3 shows the application is a typical bus network. A coupler at each node splits off part of the power from the bus and carries it to a transceiver in the attached equipment. If there are many nodes on the bus, the couplers typically have a large splitting ratio so that only a small portion of the light is tapped at each node. The throughput power at each coupler is much greater than the tap power.

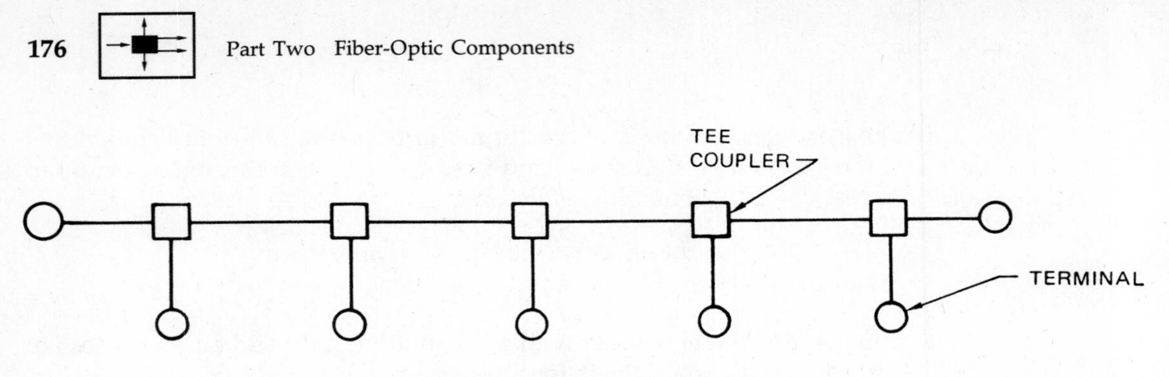

FIGURE 12–3 Tee network

Tee couplers are of greatest use when there are a few terminals on the bus. Consider a bus having N terminals. A signal must pass through $N - 1$ couplers before arriving at the receiver. In a coupler having only throughput and tap losses (i.e., no excess loss), the total distribution loss is

$$L = (N - 1)\text{loss}_{\text{THP}} + \text{loss}_{\text{TAP}}$$

Loss increases linearly with the number of terminals on the bus.

Unfortunately, we must also account for excess loss and connector loss L_C (including any loss from diameter and NA mismatches) at each coupler. Since a connector is required at both the input and output ports of the coupler, the number of required connectors is $2N$. These losses also add linearly with the number of terminals, so the real total distribution loss becomes

$$L = (N - 1)\text{loss}_{\text{THP}} + \text{loss}_{\text{TAP}} + 2NL_C + \text{loss}_{\text{E}}$$

As the number of terminals added to a network using tee couplers increases, losses mount quickly. As a result, tee couplers are only useful when a small number of terminals are involved. The difference in losses between an ideal network (having only throughput and tap losses) and a real network (having also excess and connector loss) rapidly becomes greater.

The network in Figure 12–3 is one directional. A transmitter at one end of the bus communicates with a receiver at the other end. Each terminal also contains a receiver. A duplex network can be obtained by adding a second fiber bus. It can also be obtained by using an additional directional coupler at each end and at each terminal. Such additions allow signals to flow in both directions. Figure 12–4 shows an example.

The loss characteristics of a tee network require receivers to have large dynamic ranges. Consider a 10-terminal network with terminal 1 serving as the transmitter. The power received at terminal 2 and at terminal 10 will be quite significantly different. Thus, the dynamic range of the receiver in each terminal must be large. If we assume a coupler with a 10:1 splitting ratio, assume excess and connector

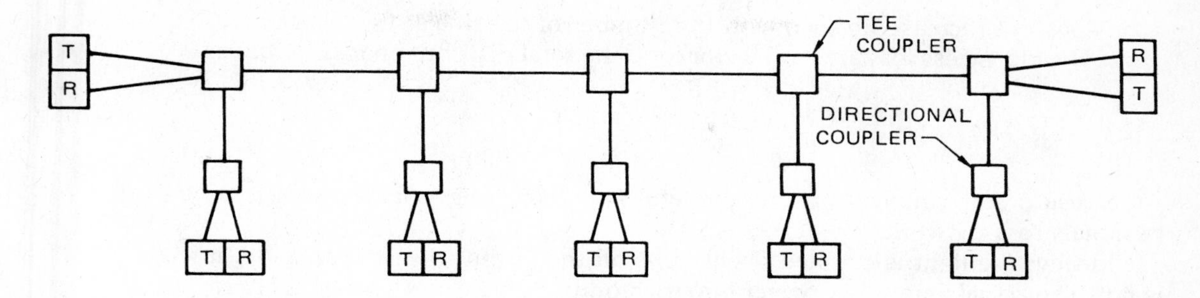

FIGURE 12–4 Tee network with directional coupler

losses at each coupler total 2 dB, and ignore all other losses in the system, the difference between receiver power at terminal 1 and at terminal 10 is about 30 dB. The receivers require a dynamic range of 30 dB or better. Other real losses will, of course, increase the requirement.

Notice that failure of a single coupler does not shut down the entire network. It simply divides the network into two smaller networks, one on each side of the failed coupler.

STAR COUPLER

The *star coupler* is an alternative to the tee coupler that makes up for many of the latter's drawbacks. Figure 12–5 shows a transmissive star coupler that has an equal number of input and output ports. For a network of N terminals, the star coupler has $2N$ ports. Light into any input port is equally divided among all the outport ports.

The insertion loss of a star coupler is the ratio of the power appearing at a given output port to that of an input port. Thus the insertion loss varies inversely with the number of terminals:

$$\text{loss}_{\text{IN}} = 10 \log_{10} \frac{1}{N}$$

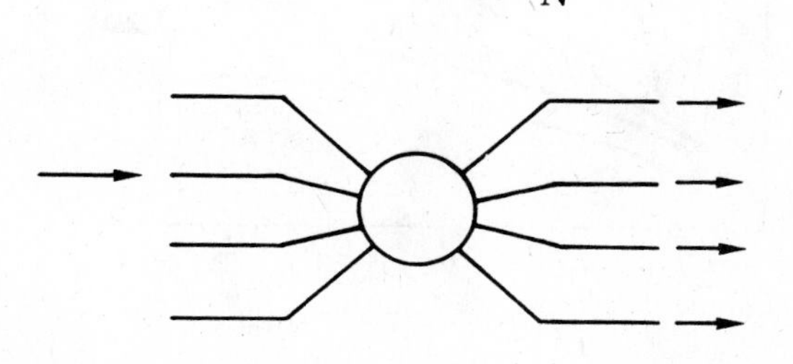

FIGURE 12–5 Star coupler

Loss does not increase linearly with the number of terminals.

If we add excess loss and connector loss, the total distribution loss becomes

$$L = 10 \log_{10} \frac{1}{N} + \text{loss}_E + 2L_C$$

As a result, star couplers are more useful for connecting a large number of terminals to a network.

Ideally, the light is coupled evenly into all the output ports. Practically, it is not. The actual amount of power into each output port varies somewhat from the ideal determined from the insertion loss. *Uniformity* is the term used to specify the variation in output power at an output port. Uniformity is expressed either as a percentage or in decibels. Consider a coupler in which the output power at each port is 50μW. A uniformity of ±0.5 dB means that the actual power will be between 45 and 56 μW. If the uniformity figure increases to ± dB, the output varies between 40 and 63 μW.

Figure 12–6 compares loss with the number of terminals for ideal and real tee and star couplers. The figure assumes an excess loss of 1 dB and a connector loss of 1 dB for each connector, for a total of 3 dB. The figure illustrates the important difference between tee and star couplers. Excess and connector losses occur at each coupler. Since a star network requires only one coupler, these losses occur only once. With a tee network, they increase linearly with the number of

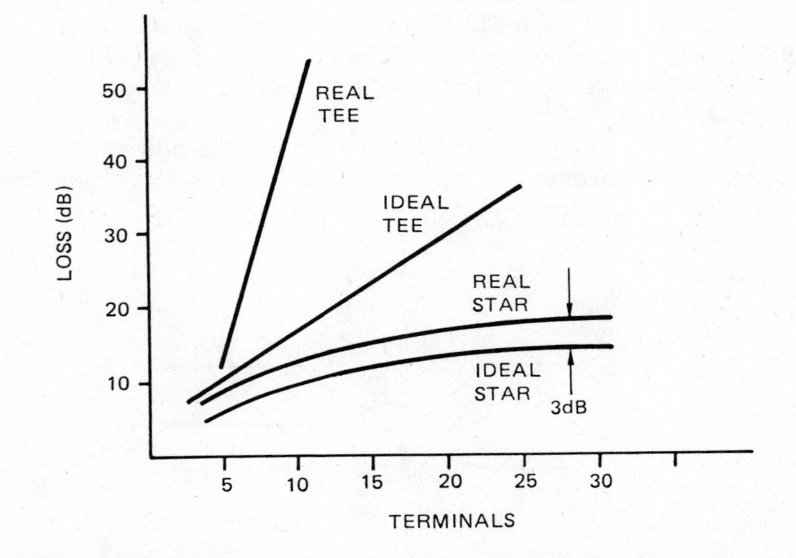

FIGURE 12–6 Loss versus number of terminals for tee and star couplers

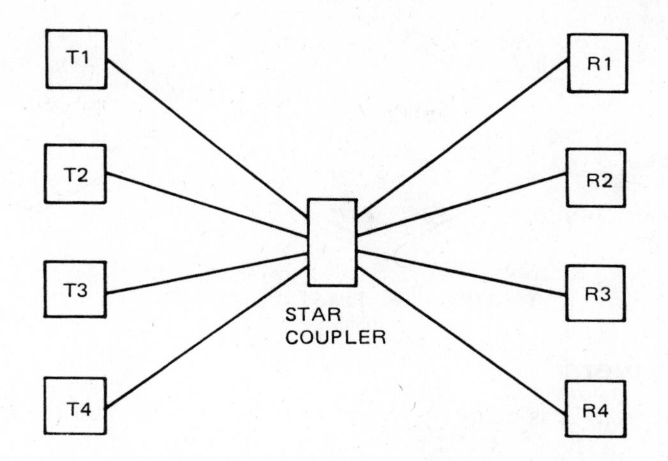

FIGURE 12–7 Star coupler application

couplers in the system. Loss in a star network increases much slower than in a tee network as the number of terminal increases.

One reason for using a tee-coupled network instead of a star-coupled network is that the tee network requires less cable. The centralized location of a star coupler requires significantly more cable to connect widely separated terminals.

Figure 12–7 shows a block diagram of a typical star network using a 4×4 star coupler.

REFLECTIVE STAR COUPLERS

A *reflective star coupler* contains *N* number of ports in which each port can serve as an input or an output. Light injected into any one port will appear at all other ports.

COUPLER MECHANISMS

This section looks at several means of constructing a coupler. Most couplers are ''black boxes,'' in which the internal coupling mechanisms are of little interest to the user. Typical input/output ports to the coupler are either fiber pigtails or connector bushings. The advantage of the pigtail is that it permits greater flexibility of application, since connection to the pigtails can be made with any compatible splice or connector. Couplers using connectorized ports are naturally limited to interconnection by specific connector type. The connectorized couplers are, on the other hand, ready to go; inexperienced users can simply connect or disconnect the fiber to the coupler.

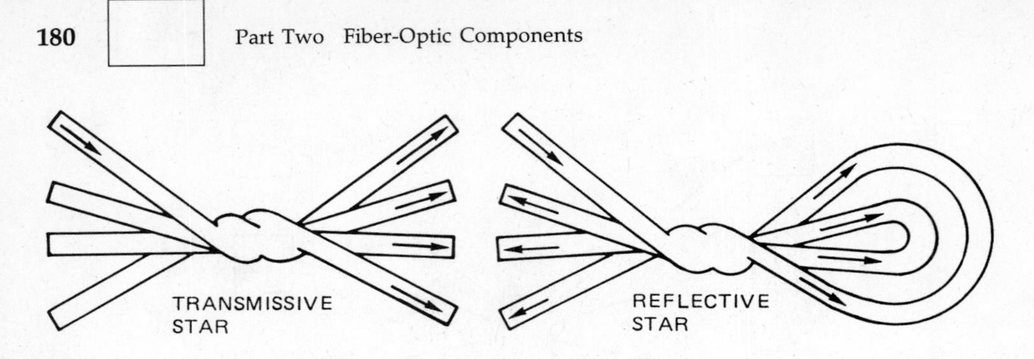

FIGURE 12–8 Fused star couplers

Fused Couplers

A *fused star coupler* is made by wrapping fibers together at a central point and heating the point. The glass will melt into a unified mass so that light from any single fiber passing through the fused point will enter all the fibers on the other side. A transmissive star coupler results when an end of each fiber is on each side of the fused section. A reflective star results when the fibers loop back so that each fiber is twice fused in the fused section. Both types are shown in Figure 12–8.

Depending on how the fibers are heated and pulled during fusing, the optical energy can be split evenly or unevenly among the fibers. Fused couplers are very small, since the fusing region is only about a tenth of an inch. They also offer very good uniformity.

Centro-Symmetrical Reflective Couplers

The basis of CSR technology is a concave mirror. Consider two fibers placed at positions equidistant from the center of curvature of a spherical mirror, as shown in Figure 12–9. When light emerges from one fiber, it spreads out as it reaches the mirror and reflects. The reflected light converges, entering the second fiber. The reflected cone is a 1:1 image of the incident cone, a mirror image on the opposite side of the mirror's centerpoint. Thus derives the term *centro-symmetric reflection*—the incident and reflected light are symmetrical to the center of curvature.

When the mirror is pivoted, the curvature changes and so does the path of the light. If we add a third fiber (Figure 12–10), the light from the input fiber can be directed to either of the output fibers, depending on the position of the mirror. Essentially, we have a switch that allows the light to be coupled into either of two fibers.

Several factors influence the performance of a CSR-based device:

- Radius of the mirror
- Refractive index of the medium between the fiber and mirror
- Fiber core diameter and NA
- Fiber angle with respect to mirror axis
- Fiber axis separation

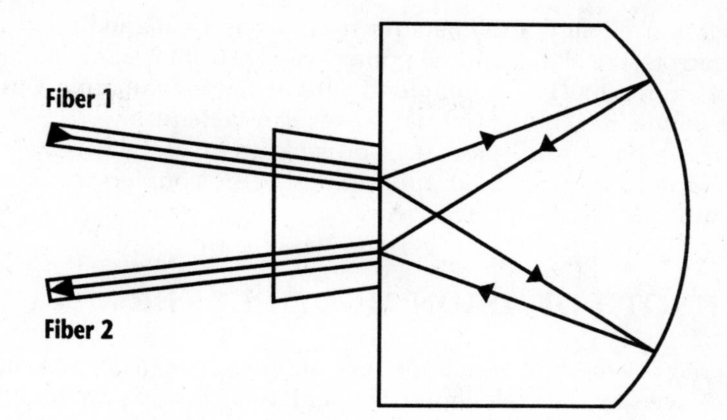

FIGURE 12–9 Principle of centro-symmetrical reflective optics (Illustration courtesy of AMP Incorporated)

Whereas the mirror radius can be optimized for the fiber being used, application flexibility requires a compromise radius to handle a variety of fiber sizes and NAs. A 9.2-mm radius yields acceptable results with common multimode core sizes (50 to 100 μm) in a parallel array. Angling the fibers preserves the symmetry of the input and output beams and reduce loss, allowing a mirror radius

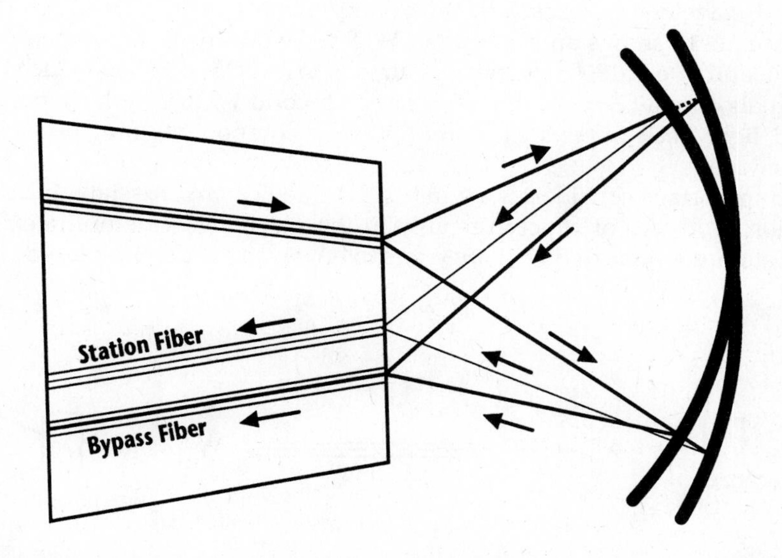

FIGURE 12–10 Pivoting the mirror allows the output fiber to be selected (Illustration courtesy of AMP Incorporated)

as small as 5 mm. Finally, the fibers must be precisely aligned with the center axis of the mirror—the allowed misalignment ranges from 0.4 μm for single mode fibers to 2.7 μm for 100/140 multimode fibers. If these competing variables are maintained within closely held limits, losses can be kept low.

Given these basic principles, it is possible to create a variety of devices necessary for flexible fiber-optic applications. CSR couplers offer very high directivity and can handle all fiber types.

WAVELENGTH-DIVISION MULTIPLEXER

Multiplexing is a method of sending several signals over a line simultaneously. In Chapter 2, we saw how telephone companies use time-division multiplexing to send hundreds of telephone calls over a single line. Each voice is first transformed in digital data by pulse-coded modulation (PCM). Each PCM-encoded signal is then allotted specific time slots within the transmission. Actual modulation and demodulation are accomplished electrically, before being presented to the fiber-optic transmitter and after being received from the fiber-optic receiver.

Wavelength-division multiplexing (WDM) uses different wavelengths to multiplex two or more signals. Transmitters operating at different wavelengths can each inject their optical signals into an optical fiber. At the other end of the link, the signals can again be discriminated and separated by wavelength. A WDM coupler serves to combine separate wavelengths onto a single fiber or to split combined wavelengths back into their component signals.

Figure 12–11 shows an example of WDM. Two transmitters, one operating at 820 nm and one at 1300 nm, present signals to a WDM coupler, which couples both signals onto a fiber. At the other end, a second WDB coupler separates the received light back into its 820- and 1300-nm components and presents it to two receivers.

Two important considerations in a WDM device are crosstalk and channel separation. Both are of concern mainly in the receiving or demultiplexing end of the system. *Crosstalk* or directivity refers to how well the demultiplexed channels

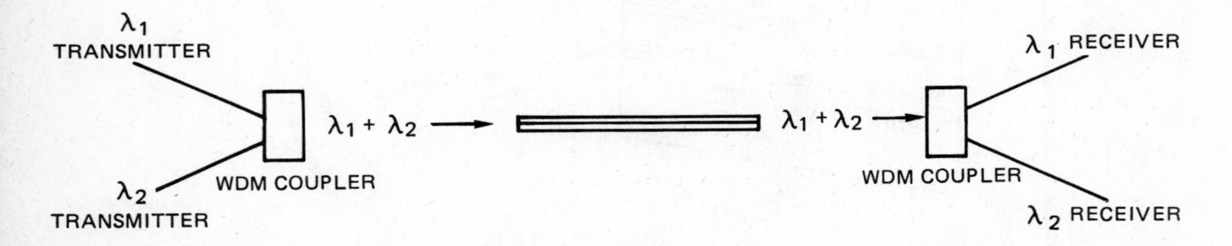

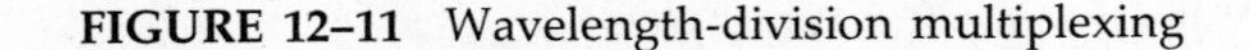
FIGURE 12–11 Wavelength-division multiplexing

are separated. Each channel should appear only at its intended port and not at any other output port. The crosstalk specification expresses how well a coupler maintains this port-to-port separation. Crosstalk, for example, measures how much of the 820-nm wavelength appears at the 1300-nm port. A crosstalk of 20 dB means that 1% of the signal appears at the unintended port.

Channel separation describes how well a coupler can distinguish wavelengths. In most couplers, the wavelengths must be widely separated, such as 820 nm and 1300 nm. Such a device will not distinguish between 1290-nm and 1310-nm signals.

Most WDM devices today operate with only a few separate optical channels. Practical, affordable couplers do not discriminate between closely separated channels. A three-port coupler, for example, might multiplex 820-nm and 1300-nm wavelengths, whereas a four-port coupler muliplexes 755-, 820-, and 1300-nm wavelengths.

The limited number of channels for WDM form the spectral requirements of a multichannel WDM device. LED sources have wide spectral widths, which require each channel to be separated widely from the others. Constructing a WDM device with many channels requires a narrow band source. Lasers, which have sufficiently narrow spectral widths, are not readily available within the range of outputs required for a high number of channels. Single-frequency lasers with a tunable wavelength have been demonstrated in the laboratory, but they are not yet commercially available.

Popular two-wavelength multiplexing schemes include:

850/1300 nm
1300/1550 nm
1480/1550 nm
985/1550 nm

WDM allows the potential information-carrying capacity of an optical fiber to be increased significantly. The bandwidth-length product used to specify the information-carrying capacity of a fiber applies only to a single channel—in other words, to a signal imposed on a single optical carrier. A fiber with a 500-MHz-km bandwidth used with a four-port WDM coupler can carry a 500-MHz signal on the 755-nm channel, a 500-MHz signal on the 820-nm channel, and a 500-MHz signal on the 1300-nm channel. Its effective information-carrying capacity has been increased to 1500 MHz.

Three approaches to WDM are diffraction gratings, interference filters, and CSR. Although there are many ways to achieve each in practical application, Figure 12–12 shows a single example each of a diffraction grating and interference filter. The *diffraction grating* is an array of fine, parallel reflecting lines spaced on the order of a wavelength. Because of the small size of the grating, the laws of reflection and refraction are replaced by those of diffraction. *Diffraction* is caused by the interaction of the wave and an object, in this case the grating. Diffraction causes deviation of waves from their paths. The grating concentrates the diffracted light to a few directions. The number of parallel lines and their spacings determine

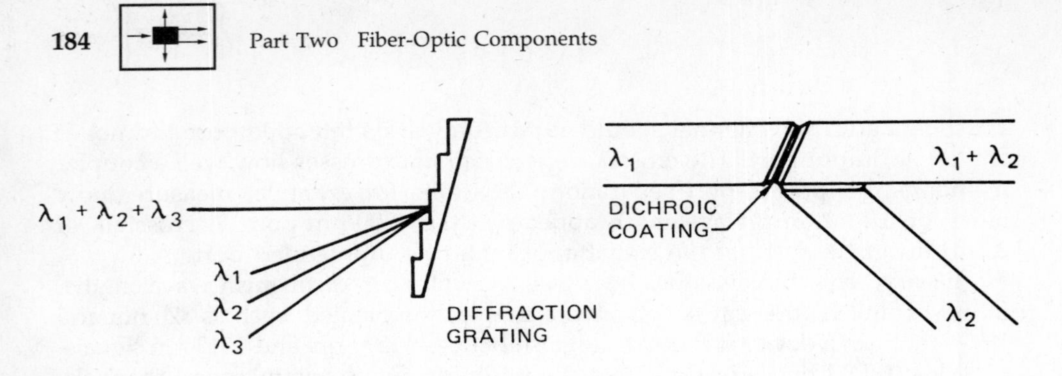

FIGURE 12–12 WDM mechanisms

the number of component channels. The grating also works in the reverse direction to combine wavelengths. The diffraction grating is most useful for systems with 5 to 10 channels. Since a correspondingly broad spectrum of sources is not available, diffraction-grating WDM is not widely available either.

An optical filter can also be used to separate wavelengths. A *dichroic* substance is one that reflects or transmits light selectively according to wavelength. The multiplexer shown in Figure 12–12 has three optical fibers arranged precisely as shown. One of the in-line fibers is coated with a dichroic substance to control the direction of light on a wavelength-sensitive basis. Although most such multiplexers will use wavelengths such as 820 nm and 1300 nm, modification of the dichroic coating permits separation of wavelengths as closely as 100 nm.

A two-wavelength CSR WDM uses two mirrors in a stacked, monolithic assembly (Figure 12–13). This assembly is packed in an aluminum tube, with the fibers all extending from one end. The fibers are mounted in a closely packed formation in the fiber holder.

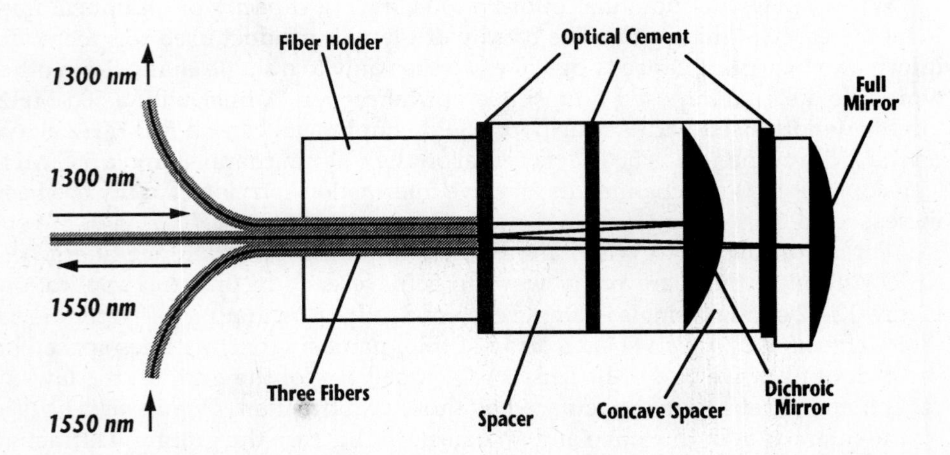

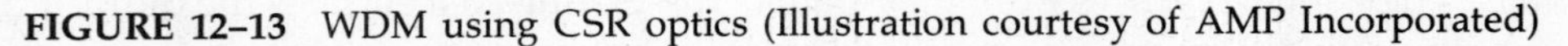
FIGURE 12–13 WDM using CSR optics (Illustration courtesy of AMP Incorporated)

The first mirror consists of a plano-convex lens with a dichroic mirror coating. It is aligned with the fiber holder and fixed in a precise position to create a 1:1 imaging between the 1300-nm short-wavelength fiber and the bidirectional transmission fiber. The dichroic coating on the lens serves as a long-wave-pass filter, reflecting the short wavelength and passing the longer 1550-nm light.

The longer wavelength strikes the second mirror, which is separately aligned with the 1550-nm fiber. The reflected light passes a second time through the first dichroic long-pass coating. This provides very high isolation—greater than 40 dB—between the two wavelengths. For example, if 100 μW of 1300-nm power is present on its fiber, only 10 nW of that power will couple onto the 1550-nm fiber. The only penalty is a 0.1 dB greater insertion loss that results from passing through the mirror twice. Because of this, the insertion loss will be less than 1 dB for one of the two wavelengths and less than 1.1 dB for the other. The device also offers excellent isolation (40 dB) and directivity (55 dB) to ensure negligible optical crosstalk.

A bidirectional application uses two different WDMs. One end uses a Type A WDM—equivalent to the one described above with a long-pass dichroic mirror on the first lens. The other end uses a Type B WDM, which is oppositely configured: the dichroic mirror is a short-pass one that reflects the 1550-nm light and passes the 1300-nm light. Figure 12-14 shows a bidirectional application, including performance characteristics for each WDM.

Insertion Loss Performance **Crosstalk (Isolation) Performance**

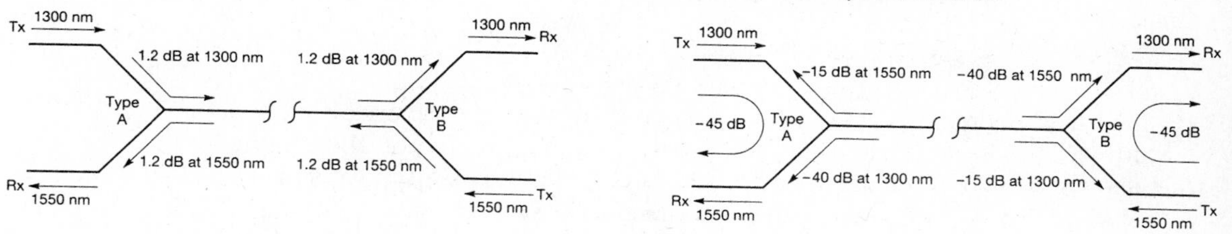

Specifications — Singlemode and Multimode

Performance Characteristics	Type A: Measured in Shorter Wavelength Band	Type A: Measured in Longer Wavelength Band	Type B: Measured in Shorter Wavelength Band	Type B: Measured in Longer Wavelength Band
Insertion Loss (Max.)	1.2 dB	1.2 dB	1.2 dB	1.2 dB
Far-end Isolation (Max.)	–40 dB	–15 dB*	–15 dB*	–40 dB
Directivity (near-end crosstalk) (Max.)	–45 dB	–45 dB	–45 dB	–45 dB
Backreflection (Max.)	–35 dB	–35 dB	–35 dB	–35 dB

*Crosstalk to opposite wavelength transmitter (negligible optical effect)

FIGURE 12–14 Bidirectional application using two CSR WDMs (Illustration courtesy of AMP Incorporated)

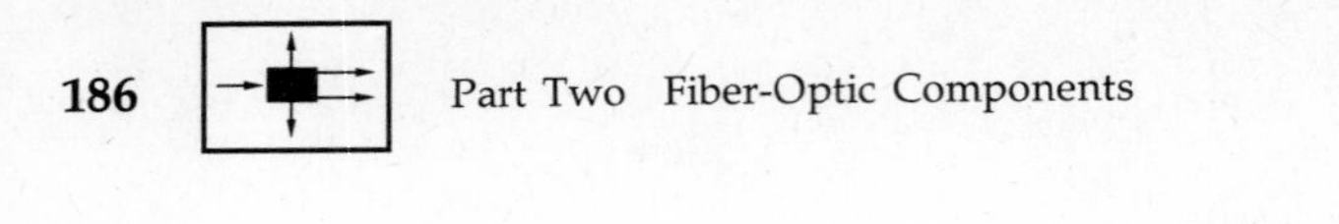

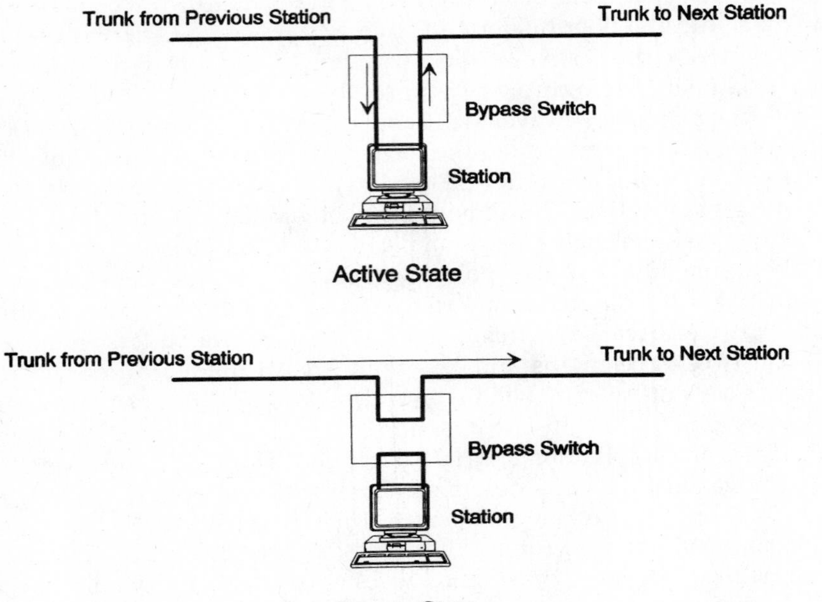

FIGURE 12–15 Function of bypass switch in a network

OPTICAL SWITCH

It is sometimes necessary to switch light between two or more fibers. A passive coupler distributes light onto all output fibers but does not allow light to be selectively switched between them. Figure 12–15 shows a typical situation of a network. Nodes (workstations) are connected in a serial, daisy chain fashion. In other words, the transmitter of one node is connected to the receiver of the next.

FDDI and other optical token-ring networks require a bypass switch to disconnect nodes from the network. When a node is active, the light passes into the node's receiver and out its transmitter. When a node is inactive, however, the incoming and outgoing lines must be optically connected by bypassing the node. If the node fails or is turned off, the network will also fail unless the node is bypassed.

An optical bypass switch fills this function. Figure 12–15 illustrates the bypass function for an FDDI network (which is discussed in detail in Chapter 15). Notice that in the bypass state, the transmitter of the station is connected to the receiver. This is useful for troubleshooting and test purposes, since the station's transceiver can operate without being attached to the network.

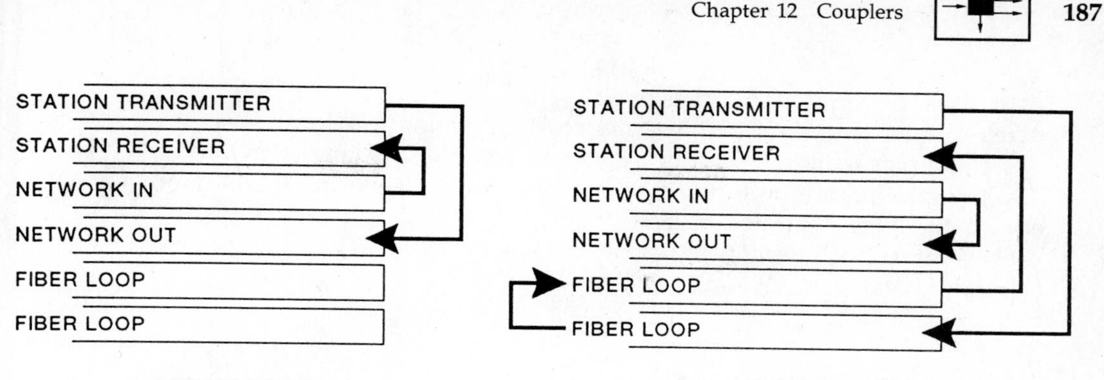

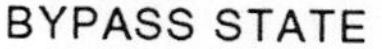

FIGURE 12–16 Function of fibers in CSR bypass switch

The basic operation of a bypass switch is the same as in Figure 12–10. The mirror pivots to direct light into different fibers. An actual switch can be somewhat more complicated in that it contains more than three fibers.

A CSR-based bypass switch uses five fibers cemented into a precision-etched fiberholder for highly accurate positioning. Figure 12–16 shows the function of each fiber and the direction of light in both the active and bypass states. (The mirror and reflections have been omitted for clarity.) Notice that one of the fibers is simply a loop that provides input and output within the switch. The loop serves to connect the station's transmitter port to its receiver port in the bypass state.

The mirror mounts on a bearing that permits it to pivot on an axis perpendicular to the plane containing the fibers. When the mirror is in one position, light from the network is reflected off the mirror into the fiber attached to the node's receiver. Similarly, light from the node's transmitter is reflected onto the network.

When the mirror is pivoted by an electromagnetic actuator to its other position, incoming light is reflected back onto the network output line directly, bypassing the node completely.

This CSR approach is highly stable, since the mirror can be pivoted to one of two highly stable positions. The fibers themselves, precisely placed during manufacture and coated with an antireflective coating to reduce back reflections, do not have to be moved.

Insertion loss for CSR switches is typically 0.6 dB and crosstalk is less than 50 dB. The switches are quite compact—a bypass switch is less than 0.5 inch high—and reliable. They exhibit environmental stability and a life in excess of 1 million switching cycles.

SUMMARY

- A coupler is used to distribute light.
- The two principal types of couplers are tee couplers and star couplers.

- A network is a multipoint fiber-optic system.
- A tee coupler is most useful in a network containing few terminals.
- A star coupler is most useful in a network containing many terminals.
- Wavelength-division multiplexing is a method of multiplexing two or more optical channels separated by wavelength.
- A fiber-optic switch permits light from one fiber to be switched between two or more fibers.

REVIEW QUESTIONS

1. Name the two main types of couplers.
2. Sketch the operation of a three-port directional coupler.
3. Explain why a tee coupler is more useful when used in a network with few terminals. Give two advantages of a tee coupler over a star coupler.
4. Describe and sketch the basis of centro-symmetrical optics.
5. Calculate the loss for a 4×4 star coupler connected to four terminals, and for an 8×8 star coupler connected to eight terminals. Ignore excess and connector losses.
6. What is the difference between transmissive and reflective star couplers?
7. Assume a fiber has a 600-MHz-km bandwidth and, through WDM, transmits optical channels at 850 nm and 1300 nm. For a 1-km link, what is the highest bandwidth of each channel? Why?
8. What is the total information-carrying capacity of the fiber described in Question 7?
9. Describe how the application of a fiber-optic switch differs from the operation of a passive coupler.

Part Three

FIBER-OPTIC SYSTEMS

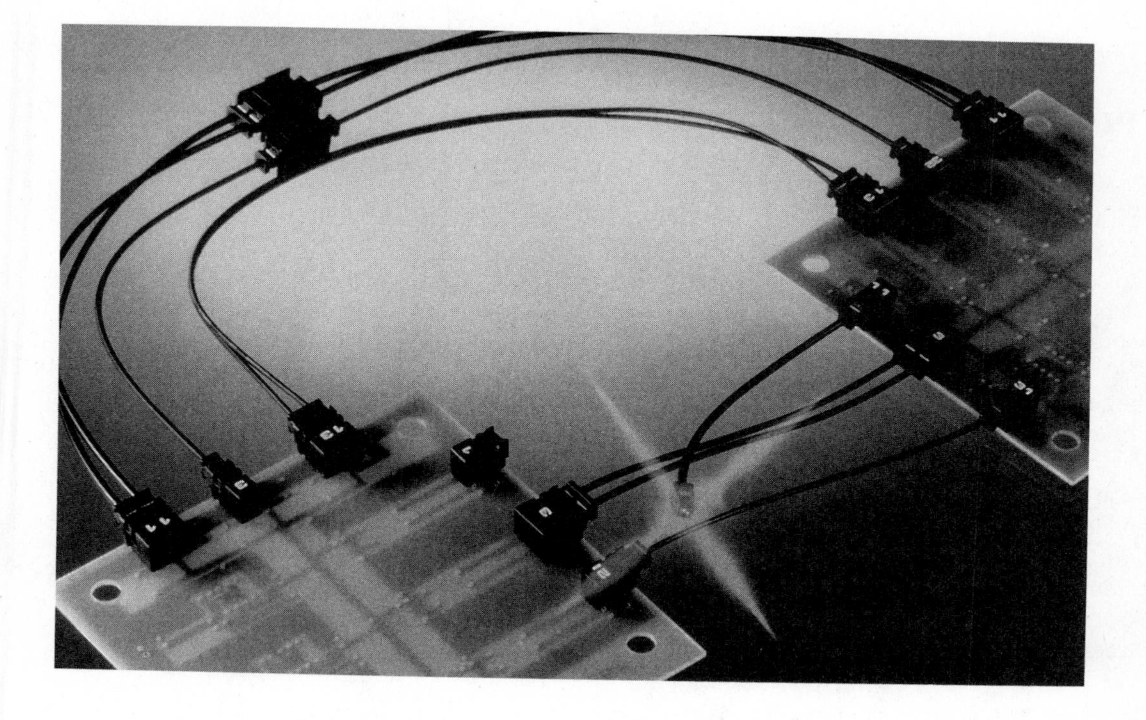

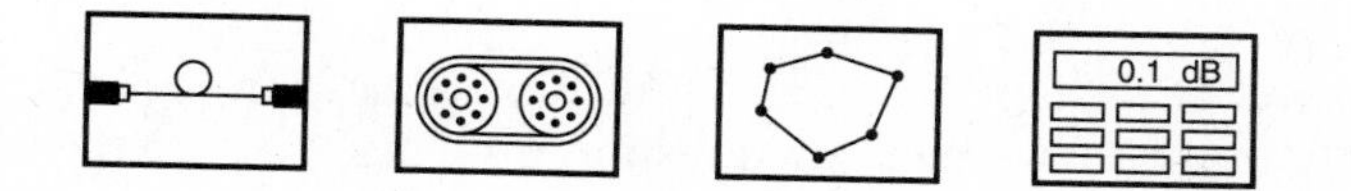

CHAPTER 13

The Fiber-Optic Link

We have now looked at the main components of a fiber-optic link, at cables, sources and transmitters, detectors and receivers, and connectors and couplers. This chapter looks at how these components are brought together into a link and at how a link is planned. It details the ingredients of a power budget and a rise-time budget, which are the two fundamental requirements that ensure that a link meets its intended application requirements. We will learn how to construct and evaluate such budgets for a simple fiber-optic system.

PRELIMINARY CONSIDERATIONS

The first step in any application is to have an application suitable to fiber optics in the first place. Wiring your doorbell with a fiber-optic link might be an interesting exercise, but it is hardly practical. Similarly, linking a personal computer to a nearby printer is most easily and inexpensively done with a standard electrical cable. If, however, the printer is a great distance away or there are strong noise sources nearby that interfere with transmission, fiber optics becomes of interest.

The decision to use fiber optics involves comparing its advantages, disadvantages, and costs against competing copper solutions. We will, however, assume fiber optics has been chosen as the best medium.

The next step is to decide whether to buy a complete fiber-optic system, to use packaged transmitters and receivers and build your own system from there, or to design and build your own transmitters and receivers as well. Because of our interests in this chapter, we select the middle choice. The third option sidetracks us from the main purpose of this chapter, which is budgeting the optical portion of the link. The first option, which will be installed by a vendor or contractor, leaves us with nothing to discuss.

SYSTEM SPECIFICATIONS

In planning a fiber-optic system, we must define our application requirements so that we can specify our needs. The main question involves the data rates and distances involved: How far? How fast? These questions give us the basic application constraints. Beyond that, we must specify the BER required.

Now that we have the main requirements—distance and data rate—we can begin to evaluate the other factors involved:

- Type of fiber
- Operating wavelength
- Transmitter power
- Source type: LED or laser
- Receiver sensitivity
- Detector type: pin diode, APD, IDP
- Modulation code
- BER
- Interface compatibility
- Number of connectors
- Number of splices
- Environmental concerns
- Mechanical concerns
- Other special concerns

The environmental and mechanical concerns involve such issues as temperature and humidity ranges, indoor/outdoor application, flammability requirements, and so forth. These factors will especially affect the choice of fiber-optic cable.

We can see that many of these questions are related and cannot be as easily separated as we have done here. The receiver sensitivity is influenced by the choice of detector. The receiver sensitivity sets the minimum optical power required by the receiver. The power arriving at the receiver, however, depends on the transmitter power and the fiber attenuation. We can specify a very sensitive receiver that will allow us to use a transmitter of lower power. Or we can use a less sensitive receiver, in which case our transmitter must be more powerful.

Consider a given transmitter, fiber-optic cable, and receiver. Now assume the power at the receiver is insufficient to meet the BER requirement of the application. We actually have five choices to remedy the situation:

1. A transmitter with a higher output power
2. A fiber with lower attenuation
3. A shorter transmission distance, to lower losses along the fiber length
4. A receiver with a lower sensitivity level
5. A modification of the system requirements, so that a lower BER is acceptable (a lower BER implies a lower minimum power at the receiver)

The point is that planning a fiber-optic link is not a cut-and-dried, step-by-step procedure. However, there are logical and rational ways to proceed. One approach is the link power budget.

POWER BUDGET

The *power budget* provides a convenient way to analyze and quantify losses in a link. The basic task of the link is to deliver enough of the transmitter's power to the receiver. The budget is the difference between the transmitter power and the receiver sensitivity. If the peak transmitter power is –10 dBm and the receiver sensitivity is –30 dBm, the power budget is 20 dB. The link cannot tolerate more than 20 dB of loss if it is to opeate properly. We must then analyze all losses within the link to ensure that their total does not exceed 20 dB. If the total loss is less than 20 dB, the remaining power is our margin. If, for example, the total loss is 14 dB, the margin is 6 dB.

A prudently planned link has a power margin. The power margin reserves power for declines in transmitter power due to aging of the source, additional losses from link repairs (as by splicing a broken fiber), and losses that might occur from cable bends or mechanical stresses. Typical margins are 3 to 6 dB.

Figure 13–1 is a graphic representation of a power budget. The graph depicts the power levels at different points in the link to finally show the power arriving at the receiver. The difference between that power and the receiver sensitivity is the margin.

Let us assume the following numerical values for the link in Figure 13–1:

Transmitter power	–10 dBm (100 μW)
Fiber attenuation	3 dB/km

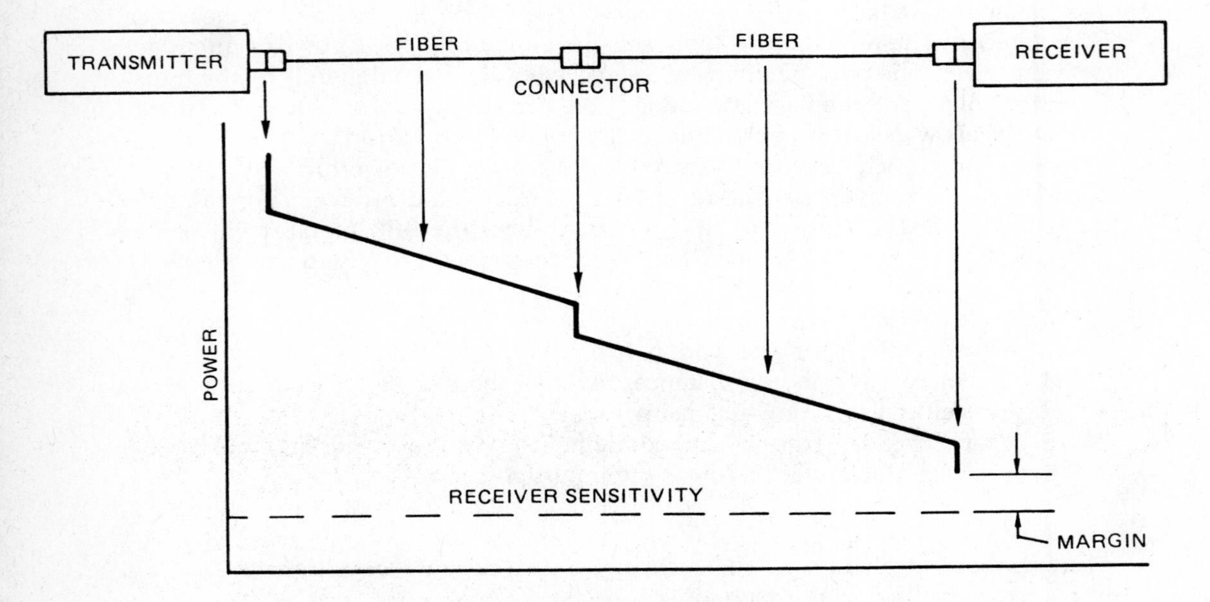

FIGURE 13–1 Link power budget

Link length	2 km
Connector loss	1 dB each interconnection
Receiver sensitivity	-30 dBm (1 μW)

The total loss from fiber attenuation is 6 dB for the 2-km run. Total loss between transmitter and receiver is 7.5 dB, leaving a margin of 12.5 dB.

A MORE COMPLEX EXAMPLE

The link power budget just shown was very straightforward. Let us take a look at a more complex example. The basic layout is the same as the link in Figure 13–1, but now we must calculate the power coupled from the transmitter into the fiber and from the fiber into the receiver. In addition, we will assume the lengths of fiber are different types of fiber. We will require a 6-dB power margin. The question is, how long can the second length of fiber be?

Here are the pertinent specifications:

Transmitter	
Output power	250 μW (–6 dBm)
Output diameter	100 μm
NA	0.30
Connector loss	1 dB
Fiber 1	
Size	85/125 μm
NA	0.26
Attenuation	5 dB/km
Length	2 km
Connector	
Loss (max)	1 dB
Fiber 2	
Size	100/140 μm
NA	.3
Attenuation	5 dB/km
Length	?
Receiver	
Sensitivity	125 nA (–39 dBm)
Diameter	150 μm
NA	0.4
Connector loss	1 dB

A brief inspection of the specifications shows many instances of diameter and NA mismatches. The example is a practical one, however. The specifications are typical of those encountered in assembling real components. They were not chosen simply to complicate the example with mismatches: The mismatches occur easily enough in the real world. The mismatches were not included in the first example because the transmitter and receiver manufacturer had already accounted for those when the device was specified.

The link power budget is 33 dB, since the difference between the –39-dBm receiver sensitivity and the –6-dBm transmitter power is 33 dB. After subtracting the 6-dB margin, we have 27 dB remaining for losses in the link. We then calculate all other losses in the link. The remaining loss is then available for fiber 2.

Transmitter Losses

Transmitter losses are from NA mismatch, diameter mismatch, and insertion loss from the connector, which is 1 dB. Loss from diameter mismatch between the 100-μm-diameter transmitter output and the 85-μm-diameter fiber is

$$\begin{aligned} \text{loss}_{\text{dia}} &= 10 \log_{10} \left(\frac{\text{dia}_{\text{fiber}}}{\text{dia}_{\text{tr}}} \right)^2 \\ &= 10 \log_{10} \left(\frac{85}{100} \right)^2 \\ &= -1.4 \text{ dB} \end{aligned}$$

Loss from NA mismatch between the 0.30 NA of the transmitter and the 0.26 of the fiber is

$$\begin{aligned} \text{loss}_{\text{NA}} &= 10 \log_{10} \left(\frac{\text{NA}_{\text{fiber}}}{\text{NA}_{\text{tr}}} \right)^2 \\ &= 10 \log_{10} \left(\frac{0.26}{0.30} \right)^2 \\ &= -1.2 \text{ dB} \end{aligned}$$

The loss at the transmitter interface is 3.6 dB. The total loss is now 3.6 dB.

Fiber 1 Loss

Since the attenuation for the 85/125 fiber is 5 dB/km, the loss for a 2-km run is 10 dB. The total loss is now 13.6 dB.

Fiber-to-Fiber Connection

Since fiber 2 has a larger core diameter and NA than fiber 1, no mismatch losses occur. The only loss is connector insertion loss of 1 dB. The total loss now is 14.6 dB.

Receiver Losses

The loss at the receiver is simply the 1 dB from the connector. Since the diameter and the NA of the detector in the receiver are larger than the diameter and NA of the fiber, no mismatch occurs. The total loss is now 15.6 dB.

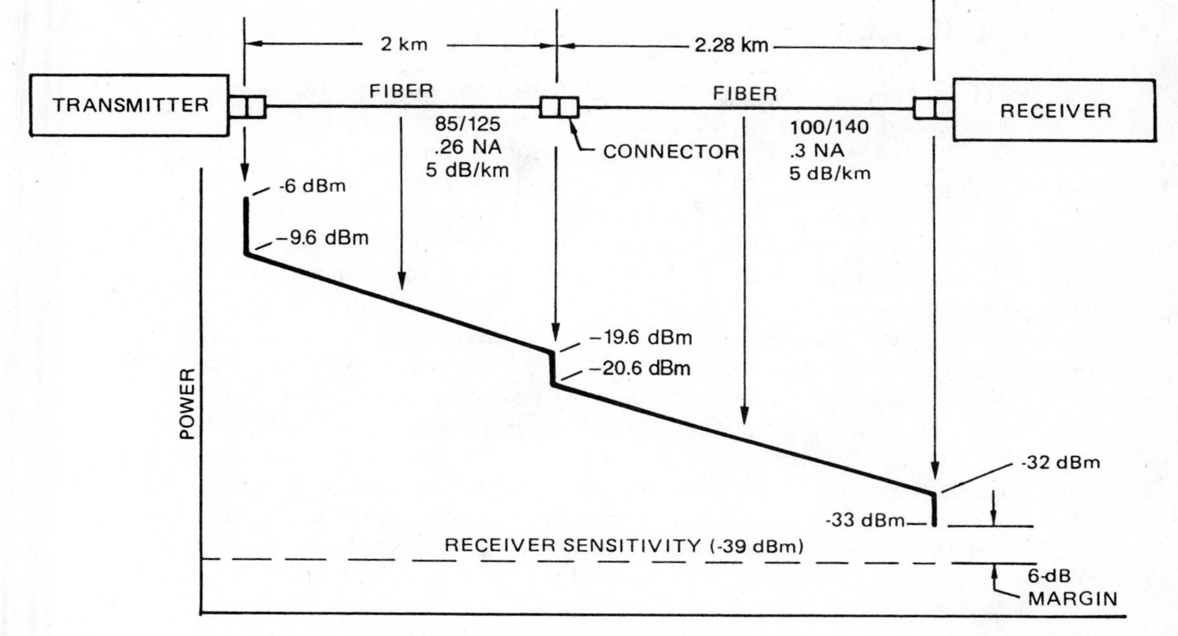

FIGURE 13–2 Link power budget example

Fiber 2 Loss

The losses calculated so far total 15.6 dB. Since allowed losses were 27 dB, we have 11.4 dB left for attenuation in the 100/140 fiber. This fiber has an attenuation of 5 dB/km, so we can run it 2.28 km. Figure 13–2 shows this application, and the losses and power levels at each point in the system.

If, in this example, the length of fiber 2 was set at 0.5 km, we would have 8.6 dB let in our budget. We could reduce the power output of the transmitter to extend the life of the source. Or we could operate the receiver above the required sensitivity to obtain a better BER.

The budget can be written in tabular form as shown in Figure 13–3.

Transmitter power	–6 dBm	
Receiver sensitivity	–39 dBm	
Power budget		33 dB
Transmitter losses	3.6 dB	
Fiber 1 loss (2 km)	10 dB	
Fiber-to-fiber loss	1 dB	
Fiber 2 loss (2.28 km)	11.4 dB	
Receiver loss	1 dB	
Total loss		27 dB
Power margin		6 dB

FIGURE 13–3 Power budget in tabular form

ADDED COMPLEXITIES

A power budget is complicated by the facts of the real world. Many components have a range within their specifications. The peak output of a transmitter may be specified with a 3-dB difference between its minimum and its maximum. The connector loss used was a 1-dB maximum. A typical value might be closer to 0.6 dB, and there is a possibility it might reach 0.2 dB. Fiber attenuation varies somewhat with temperature, usually decreasing from that specified. Therefore, a link budget should be calculated for minimum and maximum values to obtain worst-case and best-case values.

In some cases, a link that operates satisfactorily under worst-case conditions of minimum transmitter peak power and maximum losses could overdrive the receiver under best-case conditions of maximum transmitter peak power and minimum losses. Such problems most likely will occur if the receiver has a small dynamic range. If the receiver becomes overdriven, the drive current to the transmitter can be reduced, or additional attenuation can be introduced into the system. Connectors used in the telephone industry, for example, have barrels that fit over them. The sole purpose of the barrel is to provide additional attenuation. These barrels are added as needed.

RISE-TIME BUDGET

The power budget analysis ensures that sufficient power is available throughout the link to meet application demands. Power is one part of the link requirement. The other part is bandwidth or rise time. All components in the link must operate fast enough to meet the bandwidth requirements of the application. The rise-time budget allows such analysis to be performed.

As we saw in the chapters on sources and detectors, active devices have finite response times in response to inputs. The sources and detectors do not turn on and off instantaneously. Rise and fall times determine response time and the resulting bandwidth or operating speed of the devices.

Similarly, dispersion limits the bandwidth in the fiber. In a single-mode fiber, only material dispersion from the source spectral width severely limits bandwidth. In a multimode fiber, the delay caused by light traveling in different modes also limits bandwidth or cable rise time.

When the bandwidth of a component is specified, its rise time can be approximated from

$$t_r = \frac{0.35}{\text{BW}}$$

This equation accounts for modal dispersion in a fiber. The rise time must be scaled to the fiber length of the application. If the cable is specified at 600 MHz/km, and the application is 2 km, the bandwidth is 300 MHz and the rise time is 1.6 ns.

In addition, the rise-time budget must include the rise times of the transmitter and receiver, which are typically specified for packaged devices. Transmitter and receiver rise times are used instead of source and detector rise times, since the transmitter and receiver circuits will limit the maximum speed at which opto-electronic devices can operate. For the receiver, rise time/bandwidth may be limited by either the rise time of the components or by the bandwidth of the *RC* time constant.

Connectors, splices, and couplers usually do not affect system speed, and they do not have to be accounted for in the rise-time budget.

When all individual rise times have been found, the system rise time is calculated from the following:

$$t_{rsys} = 1.1\sqrt{t_{r1}^2 + t_{r2}^2 + \cdots + t_{rn}^2}$$

The 1.1 allows for a 10% degradation factor in the system rise time. In a typical application, the rise-time budget can also be used to set the rise time of any individual components, such as the transmitter or cable, by rearranging the equation to solve for the unknown rise time.

When rearranging the equation to solve for an individual rise time, the 1.1 factor is not used. Since, after all, the factor allows for degradation within the entire system, it should be applied to individual components.

Consider as an example a 20-MHz application operating over 2 km. The fiber used has a 400-MHz-km bandwidth. The receiver rise time is 10 ns. What is the rise time required of the transmitter?

The required system rise time is 17.5 ns. Fiber rise time is 1.75 ns. Solving for the transmitter rise time gives

$$\begin{aligned} t_{rtrans} &= \sqrt{17.5^2 - 10^2 - 1.75^2} \\ &= 14.25 \text{ ns} \end{aligned}$$

The transmitter must have a rise time of about 14 ns. If we select a transmitter with a rise time of 10 ns, we will meet the rise time or bandwidth requirements of the application. With a 10-ns transmitter, the system rise time becomes

$$\begin{aligned} t_{rsys} &= 1.1\sqrt{10^2 + 1.75^2 + 10^2} \\ &= 15.7 \text{ ns} \end{aligned}$$

which is within the required 17.5 ns.

SUMMARY

- A power budget ensures that losses are low enough in a link to deliver the required power to the receiver.

- A rise-time budget ensures that all components meet the bandwidth/rise-time requirements of the link.

? REVIEW QUESTIONS

1. Using the transmitter and receiver specifications in Table 10–1 and performing a power budget analysis, determine the longest length of fiber that can be used. The fiber used is a 50/125 fiber with a bandwidth of 600 MHz and loss of 4 dB/km. Assume a 6-dB margin. Assume that a fiber-to-fiber connection with a 1-dB loss will be required after every 2 km. Use the maximum output value for the transmitter and the minimum value for the receiver.
2. Sketch a power budget graph for Question 1.
3. Repeat the analysis from Question 1 for a 100/140 fiber with a bandwidth of 400 MHz and loss of 6 dB/km.
4. Sketch a power budget for Question 3.
5. Determine the system rise time for the link analyzed in Question 1. Use the length of fiber determined from the power budget. Use the maximum rise-time values for the transmitter and receiver.

Fiber Optic Cable Installation and Hardware

This chapter looks at some of the factors involved in installing a fiber-optic cable system. These factors are of interest because they not only demonstrate the practical importance of the mechanical properties of a cable, but they also demonstrate the practical aspects of dealing with fiber optics. We will also describe some types of common fiber-optic hardware, such as closure/organizers, rack boxes, and distribution panels. This hardware is an important part of more complex fiber-optic systems.

BEND RADIUS AND TENSILE RATING

Because of their light weight and extreme flexibility, fiber-optic cables are often more easily installed than their copper counterparts. They can be more easily handled and pulled over greater distances.

Minimum bend radius and *maximum tensile rating* are the critical specifications for any fiber-optic cable installation, both during the installing process and during the installed life of the cable. Careful planning of the installation layout will ensure that the specifications are not exceeded. In addition, the installation process itself must be carefully planned. This chapter discusses the factors involved in the planning.

The minimum bend radius and maximum tensile loading allowed on a cable differ during and after installation. An increasing tensile load causes a reversible attenuation increase, an irreversible attenuation increase, and, finally, cracking of the fiber. The tensile loading allowed during installation is higher than that allowed after installation. Care must be taken in either case not to exceed the specified limits.

The minimum bend radius allowed during installation is larger than the bend radius allowed after installation. One reason is that the allowed bend radius increases with tensile loading. Since the fiber is under load during installation, the minimum bend radius must be larger. The allowed bend radius after installation depends on the tensile load.

Figure 14–1 shows the cross section of a simplex fiber and a duplex fiber, which we will use as examples in this chapter, and gives the specifications for their minimum bend radius and maximum tensile rating.

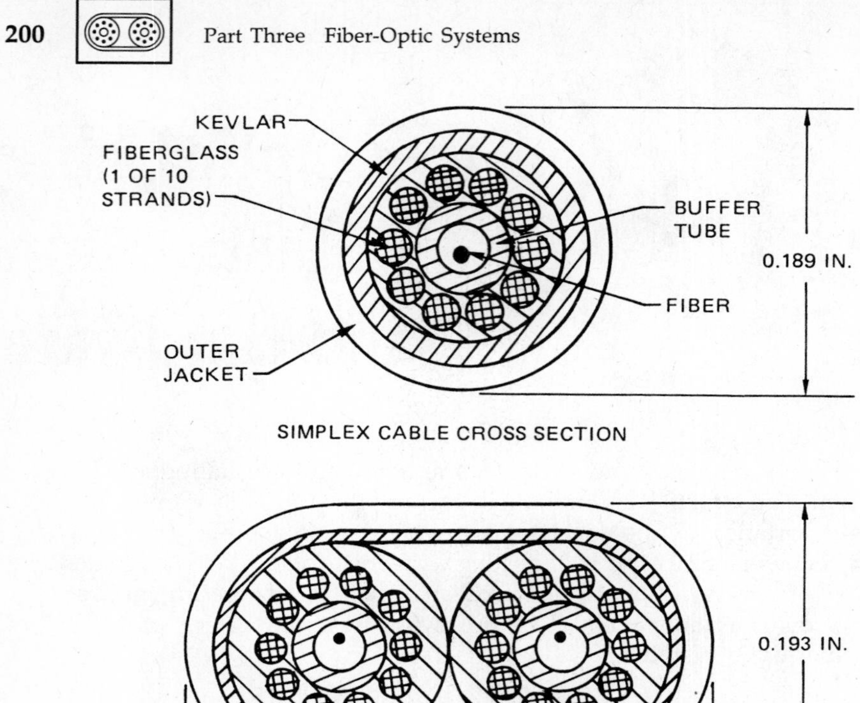

TENSILE LOADING (MAX)	CABLE
DURING AFTER	90 lb (400 N) 11.25 lb (50 N)
BEND RADIUS (MIN)	CABLE
DURING	5.9 in (150 mm)
AFTER (NO LOAD) AFTER (FULL LOAD)	1.2 in (30 mm) 4.0 in (130 mm)

FIGURE 14–1 Bend radius and tensile strength loading for cable examples (Courtesy of Canoga Data Systems)

As discussed in Chapter 7, outdoor cables are commonly multifiber cables with constructions more complex than indoor cables. Indoor cables are typically simplex or duplex.

DIRECT BURIAL INSTALLATION

Cables can be buried directly in the ground by either plowing or trenching methods. The plowing method uses a cable-laying plow, which opens the ground, lays the cable, and covers the cable in a single operation. In the trench method, a trench is dug with a machine such as a backhoe, the cable is laid, and the trench is filled. The trench method is more suited to short-distance installations.

Buried cables must be protected against frost, water seepage, attack by burrowing and gnawing animals, and mechanical stresses that could result from earth movements. Armored cables specially designed for burial are available. Cables should be buried at least 30 in. deep so they are below the frost line. Other buried cables should be enclosed in sturdy polyurethane or PVC pipes. The pipes should have an inside diameter several times the outside diameter of the cable to protect against earth movements. An excess length of cable in the pipe prevents tensile loads being placed on the cable.

AERIAL INSTALLATION

Aerial installation includes stringing cables between telephone poles or along power lines. Unlike copper cables, fiber-optic cables run along power lines with no danger of inductive interference.

Aerial cables must be able to withstand the forces of high winds, storms, ice loading, and so forth. Self-supporting aerial cables can be strung directly from pole to pole. Other cables must be lashed to a high-strength steel wire, which provides the necessary support. The use of a separate support structure is the usual preferred method.

INDOOR INSTALLATION

Most indoor cables must be placed in conduits or trays. Since standard fiber-optic cables are electrically nonconductive, they may be placed in the same ducts as high-voltage cables without the special insulation required by copper wire. Many cables cannot, however, be placed inside air conditioning or ventilation ducts for the same reason that PVC-insulated wire should not be placed in these areas: A fire inside these ducts could cause the outer jacket to burn and produce toxic gases.

Plenum cables, however, can be placed in any plenum area within a building without special restrictions. The material used in these cables does not produce toxic fumes.

TRAY AND DUCT INSTALLATIONS

The first mechanical property of the cable that must be considered in planning an installation is the outside diameter of the cable itself and the connectors. Here we discuss the cables from Figure 14–1 terminated with SMA connectors. The connector's outside diameter is 0.38 in. (9.7 mm). The outside diameter of the simplex cable is 0.189 (4.8 mm). The duplex cable has an oval cross section, 0.193 × 0.335 in. (4.9 × 8.5 mm). If the cable must be pulled through a conduit or duct, the minimum cross-sectional area required is 0.79 × 0.43 in. (20 mm × 11 mm) for duplex cable. For simplex cable, the minimum cross-sectional area is determined by the pulling grip used.

The primary consideration in selecting a route for fiber-optic cable through trays and ducts is to avoid potential cutting edges and sharp bends. Areas where particular caution must be taken are corners and exit slots in sides of trays (Figure 14–2).

If a fiber-optic cable is in the same tray or duct with very large, heavy electrical cables, care must be taken to avoid placing excessive crushing forces on the fiber-optic cable, particularly where the heavy cables cross over the fiber-optic cable (Figure 14–3). In general, cables in trays and ducts are not subjected to tensile forces;

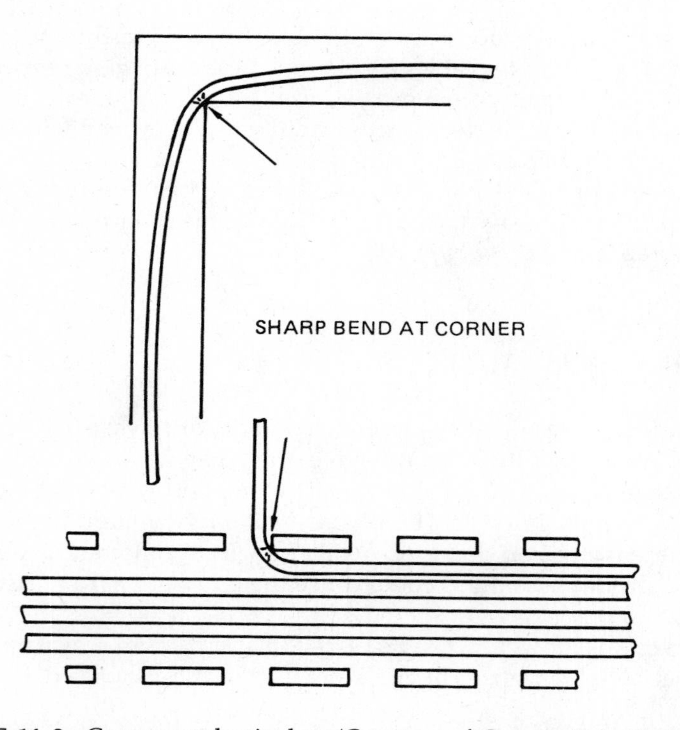

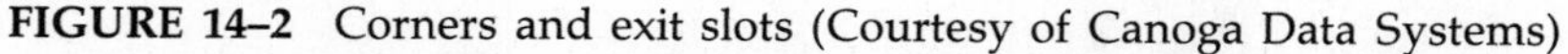
FIGURE 14–2 Corners and exit slots (Courtesy of Canoga Data Systems)

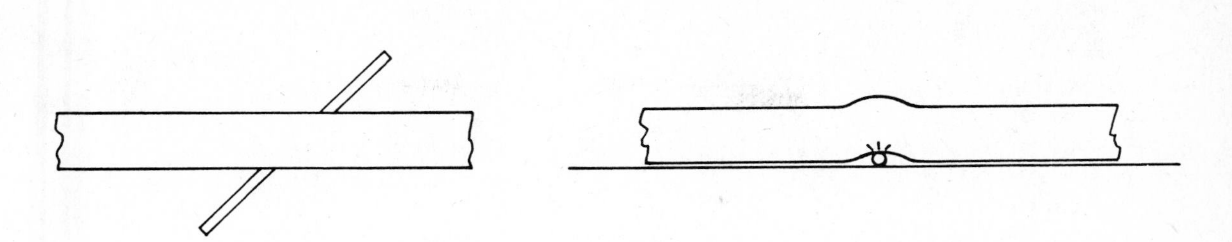

FIGURE 14–3 Crossovers (Courtesy of Canoga Data Systems)

however, it must be kept in mind that in long vertical runs, the weight of the cable itself will create a tensile load of approximately 0.16 lb/ft (0.25 N/m) for simplex cable and 0.27 lb/ft (0.44 N/m) for duplex cable. This tensile load must be considered when determining the minimum bend radius at the top of the vertical run. Long vertical runs should be clamped at intermediate points (preferably every 1 to 2 m) to prevent excessive tensile loading on the cable. The absolute maximum distance between clamping points is 330 ft (100 m) for duplex cable and 690 ft (210 m) for simplex cable. Clamping force should be no more than is necessary to prevent the possibility of slippage, and it is best determined experimentally since it is highly dependent on the type of clamping material used and the presence of surface contaminants, both on the clamp and on the jacket of the optical cable. Even so, the clamping force must not exceed 57 lb/in. (100 N/cm) and must be applied uniformly across the full-width of the cable. The clamping force should be applied over as long a length of the cable as practical, and the clamping surfaces should be made of a soft material, such as rubber or plastic.

Tensile load during vertical installation is reduced by beginning at the top and running the cable down.

CONDUIT INSTALLATIONS

Fibers are pulled through conduits by a wire or synthetic rope attached to the cable. Any pulling forces must be applied to the cable strength members and not to the fiber. For cables without connectors, pull wire can be tied to Kevlar strength members, or a pulling grip can be taped to the cable jacket or sheath. More care, as discussed later in this chapter, must be exercised with cables with connectors.

The first factor that must be considered in determining the suitability of a conduit for a fiber-optic cable is the clearance between the walls of the conduit and other cables that may be present. Sufficient clearance must be available to allow the fiber-optic cable to be pulled through without excessive friction or binding, since the maximum pulling force that can be used is 90 lb (400 N). Since minimum bend radius increases with increasing pulling force, bends in the conduit

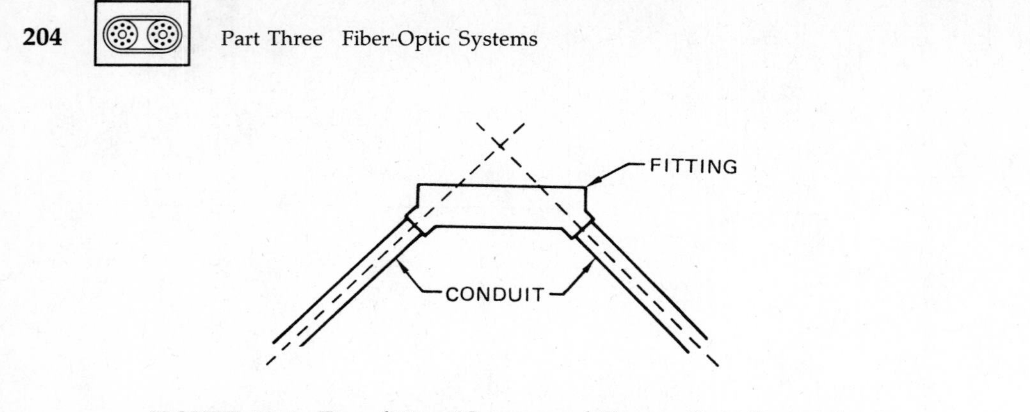

FIGURE 14–4 Turn fitting (Courtesy of Canoga Data Systems)

itself and any fittings through which the cable must be pulled should not require the cable to make a bend with a radius of less than 5.9 in. (150 mm). Fittings, in particular, should be checked carefully to ensure that they will not cause the cable to make sharp bends or be pressed against corners. If the conduit must make a 90° turn, a fitting, such as shown in Figure 14–4, must be used to allow the cable to be pulled in a straight line and to avoid sharp bends in the cable.

A pullbox is an access point in a conduit. Pullboxes should be used on straight runs at intervals of 250 to 300 ft to reduce the length of cable that must be pulled at any one time, thus reducing the pulling force. Also, pullboxes should be located in any area where the conduit makes several bends that total more than 180°. To guarantee that the cable will not be bent too tightly while pulling the slack into the pullbox, we must use a pullbox with an opening of a length equal to at least four times the minimum bend radius (4.75 in. or 120 mm). See Figure 14–5, which shows the shape of the cable as the last of the slack is pulled into the box. The tensile loading effect of vertical runs discussed in the section on tray and duct installations is also applicable to conduit installations. Since it is more difficult to properly clamp fiber-optic cables in a conduit than in a duct or tray, long vertical runs should be avoided, if possible. If a clamp is required, the best type is a fitting that grips the cable in a rubber ring. Also, the tensile load caused by the weight

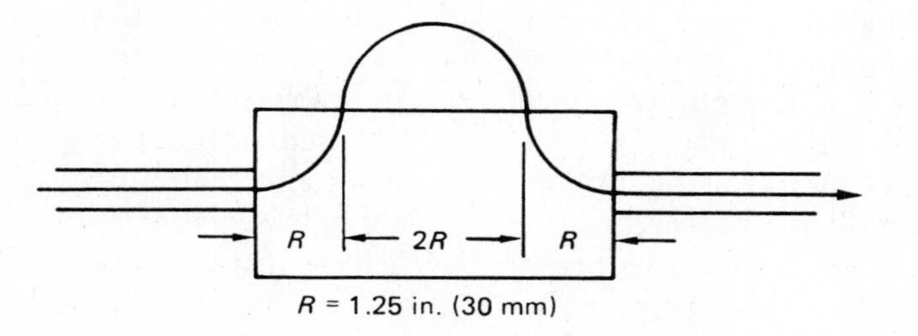

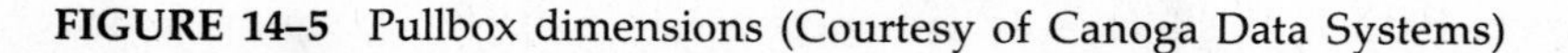

FIGURE 14–5 Pullbox dimensions (Courtesy of Canoga Data Systems)

of the cable must be considered along with the pulling force to determine the maximum total tensile load being applied to the cable.

PULLING FIBER-OPTIC CABLES

Fiber-optic cables are pulled by using many of the same tools and techniques that are used in pulling wire cables. Departures from standard methods are due to several facts: The connectors are usually preinstalled on the cable, smaller pulling forces are allowed, and there are minimum bend radius requirements.

The pull tape must be attached to the optical cable in such a way that the pulling forces are applied to the strength members of the cable (primarily the outer Kevlar layer), and the connectors are protected from damage. The recommended method of attaching a pulling tape to a simplex cable is the ''Chinese finger trap'' cable grip (Figure 14–6). The connector should be wrapped in a thin layer of foam rubber and inserted in a stiff plastic sleeve for protection. Since the smallest pulling grip available (Kellems 0333-02-044) is designed for 0.25-in. (6.4-mm)-diameter cable and the outside of the simplex cable is only 0.188 in. (4.8 mm), the cable grip should be stretched tightly and then wrapped tightly with electrical tape to provide a firm grip on the cable.

The duplex cable is supplied with Kevlar strength members extending beyond the outer jacket to provide a means of attaching the pulling tape (Figure 14–7). The Kevlar layer is epoxied to the outer jacket and inner layers to prevent inducing twisting forces while the cable is being pulled, since the Kevlar is wrapped around the inner jackets in a helical pattern. The free ends of the Kevlar fibers are inserted into a loop at the end of the pulling tape and then epoxied back to themselves. The connectors are protected by foam rubber and heat-shrink sleeving. The heat-shrink sleeving is clamped in front of the steel ring in the pulling tape to prevent pushing the connectors back toward the rest of the cable.

During pulling of the cable, pulling force should be constantly monitored by a mechanical gauge. If any increase in pulling force is noticed, pulling should immediately cease, and the cause of the increase should be determined. If the

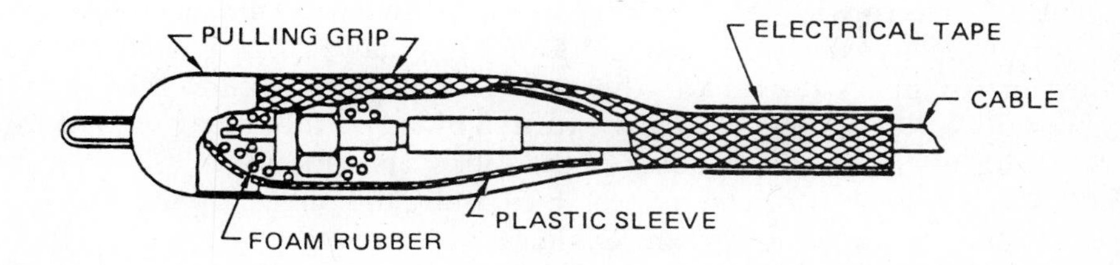

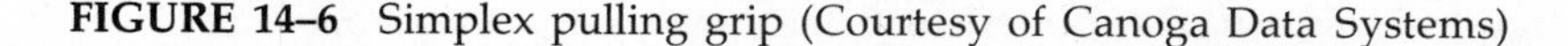
FIGURE 14–6 Simplex pulling grip (Courtesy of Canoga Data Systems)

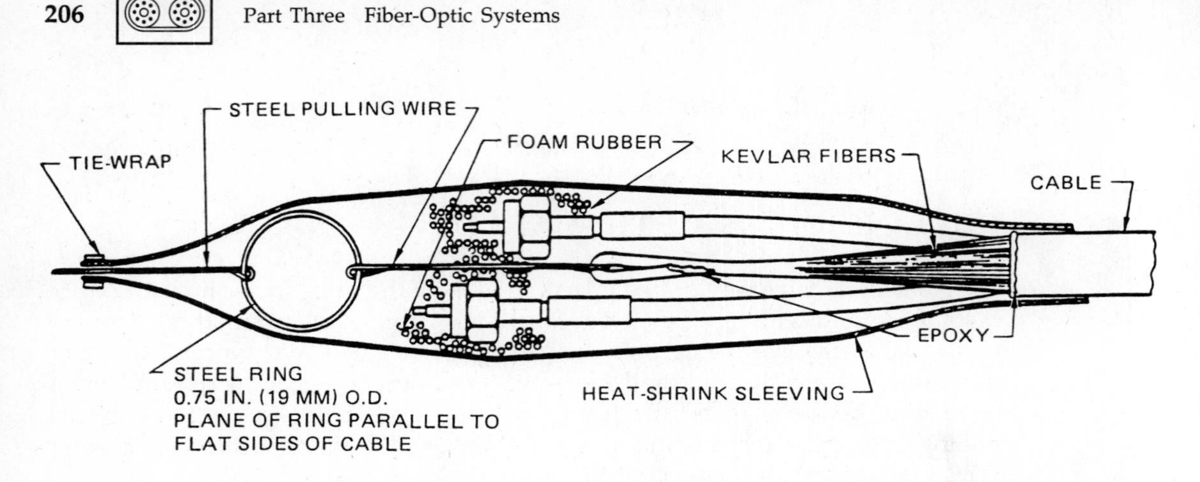

FIGURE 14–7 Duplex pulling grip (Courtesy of Canoga Data Systems)

pulling wire is subject to friction, the tensile force on the pulling wire will be more than the force applied to the fiber-optic cable, resulting in false readings.

Pull tension can be monitored by a running line tensiometer or a dynamometer and pulley arrangement. If a power winch is used to assist the pulling, a power capstan with adjustable slip clutch is recommended. The clutch, set for the maximum loading, will disengage if the set load is reached. Relying on the experience of those pulling the cable is another alternative.

If necessary during pulling-in, the cable should be continuously lubricated. The lubricant can be poured or brushed on the cable as it enters the duct or conduit. Alternatively, a slit lubrication bag can be pulled ahead of the cable so that the lubricant spills out during the pull. Lubrication should be used only for difficult pulls.

At points such as pullboxes and manholes, where the cable enters the conduit at an angle, a pulley or wheel should be used to ensure that the cable does not scrape against the end of the conduit or make sharp bends. If the portion of the cable passing over the wheel is under tension, the wheel should be at least 12 in. (300 mm) in diameter. If the cable is not under tension, the wheel should be at least 4 in. (100 mm) in diameter.

As the cable emerges from intermediate-point pullboxes, it should be coiled in a figure-8 pattern with loops at least 1 ft in diameter for simplex and duplex cable and proportionally larger diameters for larger cables. The figure-8 prevents tangling or kinking of the cable. When all the cable is coiled and the next pull is to be started, the figure-8 coil can be turned over and the cable paid out from the top. This will eliminate twisting of the cable. The amount of cable that has to be pulled at a pullbox can be reduced by starting the pull at a pullbox as close as possible to the center of the run. Cable can then be pulled from one spool at one end of the run; then the remainder of the cable can be unspooled and coiled in a figure-8 pattern and pulled to the other end of the run.

Bends in the pull should be near the beginning of the pull. Pull forces are lower if the conduit bend is near the beginning. Bends tend to multiply, rather than add, tension. For example, if a cable goes in at 20 lb, it may come out of the bend at 30 lb. If it goes in at 200 lb, it may come out at 300 lb. In the first instance, the bend added 10 lb; in the second it added 100 lb.

SPLICE CLOSURES/ORGANIZERS

A *splice closure* is a standard piece of hardware in the telephone industry for protecting cable splices. You can see them along nearly every aerial telephone run, and they are also used in underground applications. Splices are protected mechanically and environmentally within the sealed closure. The body of the closure serves to join the outer sheaths of the two cables being joined.

A standard universal telephone closure can also be used to house spliced fiber-optic cable. An *organizer panel* holds the splices. Most organizers contain provisions for securing cable strength members and for routing and securing fibers. Metal strength members may be grounded through the closure.

Organizers are specific to the splice they are designed to hold. In a typical application, the cable is removed to expose the fibers at the point it enters the closure. The length of fiber exposed is sufficient to loop it one or more times around the organizer. Such routing provides extra fiber for resplicing or rearrangement of splices.

Closures can hold one or more organizers to accommodate 12 to 144 splices.

Closures have two main applications. The first is a straight splicing of two cables when an installation span is longer than can be accommodated by a single cable. The second is to switch between types of cables for various reasons. For example, a 48-fiber cable can be brought into one end of the closure. Three 12-fiber cables, all going to different locations, can be spliced to the first cable and brought out the other end of the closure.

DISTRIBUTION HARDWARE

Within a large fiber-optic system, signals must be routed to their final destination. Such routing involves the distribution signals and the fibers that carry them. The couplers described in Chapter 12 are one example. Here we describe hardware to allow transitions from one fiber type to another or distribution of fibers to different points.

In applications such as a building, a wiring center is used as a central distribution point. From this point, fibers can be routed to their destinations. A typical application example is an outdoor multifiber cable brought into the building to the distribution point. At the wiring center, each fiber in the outdoor cable is spliced to a simplex fiber that is routed to a different location, say a different floor.

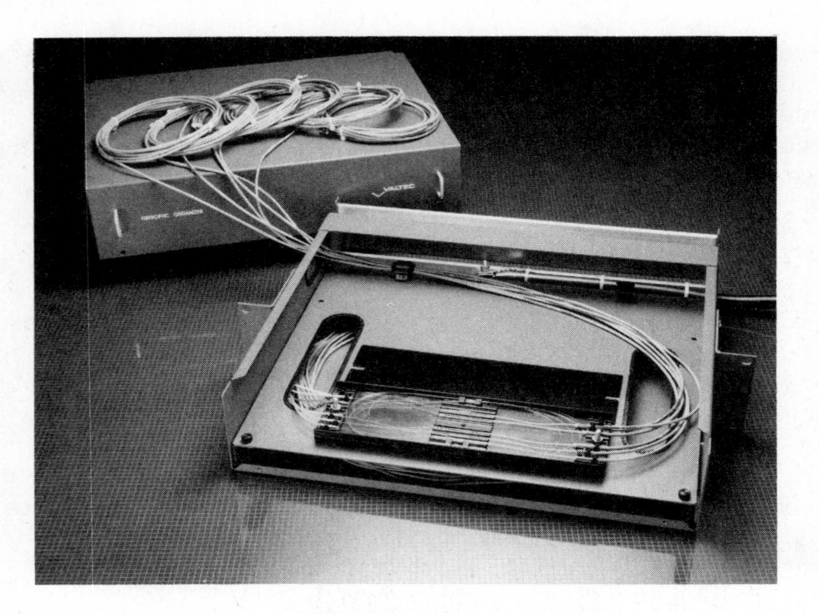

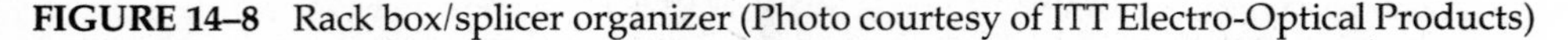

FIGURE 14–8 Rack box/splicer organizer (Photo courtesy of ITT Electro-Optical Products)

A *rack box,* such as the example shown in Figure 14–8, containing an organizer similar to the organizer used in the splice closure, serves such distribution needs. The boxes mount in 19- and 23-in. equipment racks, which are standard sizes in the telephone and electronics industries. Besides the organizer, rack boxes contain provisions for securing fibers and strain-relieving cables as well as storage room for service loops.

PATCH PANELS

Patch panels provide a convenient way to rearrange fiber connections and circuits (Figure 14–9). A simple patch panel is a metal frame containing bushings in which fiber-optic connectors plug on either side. One side of the panel is usually "fixed," meaning that the fibers are not intended to be disconnected. On the other side of the panel, fibers can be connected and disconnected to arrange the circuits as required. Like rack boxes, patch panels are compatible with 19- and 23-in. equipment racks.

Patch panels are widely used in the telephone industry to connect circuits to transmission equipment. They can also be used in the wiring center of a building to rearrange connections. The splice organizer and the patch panel serve the similar function of distribution. The difference is that the organizer is intended

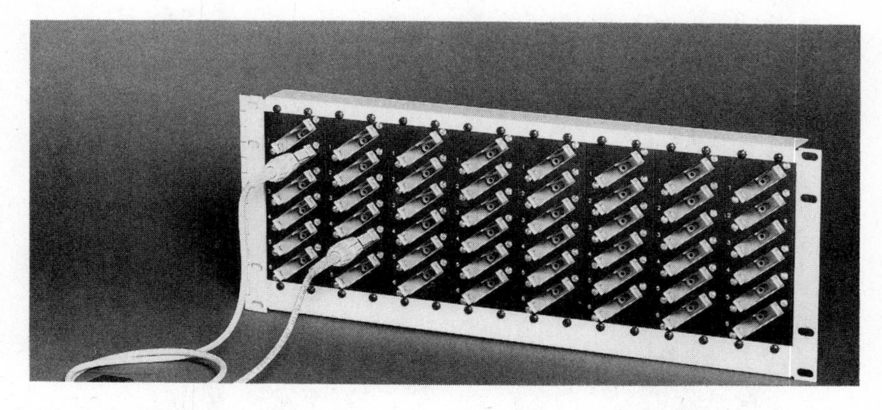

FIGURE 14–9 Patch panel (Photo courtesy of AMP Incorporated)

for fixed, unchanged connections, whereas a patch panel is used for flexible, changeable connections.

Some applications combine a splice organizer and patch panel. A large fiber is brought into the splice organizer and then spliced to other cables running directly into the "fixed" side of a patch panel. From the panel, cables can then be routed as necessary.

WALL OUTLETS

The *wall outlets* described in this section can be thought of as being similar to electrical outlets. A home or office is wired for electrical power with cable running between walls or under floors. Electrical equipment plugs into this power at a wall outlet containing an electrical outlet.

Fiber-optic wall outlets serve a similar function except that they allow connection to cables carrying optical signals. In a building or office wired with fiber optics, the outlet serves as a transition point between the building cabling and the equipment. A short simplex or duplex cable, called a *jumper* or *drop cable,* runs from the wall outlets to the equipment being served.

Figure 14–10 shows typical wall outlets, which mount in a wall in the same manner as an electrical box.

The distribution hardware clearly allows flexible, sophisticated application of fiber optics. It demonstrates that application of fiber-optic systems goes far beyond the simple point-to-point links emphasized in this book. The distribution hardware, pioneered originally for use in the telephone industry, has application outside of simple telephony. As we mentioned in Chapter 2, the information revolution has blurred the distinctions between simple telephone applications, telecommunications, and computers. Outside of telephone offices, the hardware

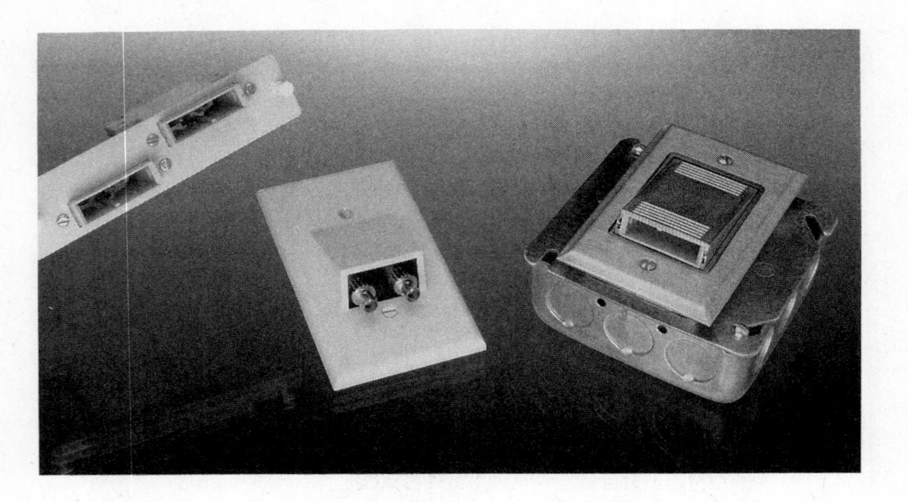

FIGURE 14–10 Wall outlets (Photo courtesy of AMP Incorporated)

finds use in wired buildings and offices, whether the application is telecommunications, computer networks, or local area networks. The couplers described in Chapter 12 are used in LANs in an office environment, for example, whereas the distribution hardware discussed here might be used to connect several LANs within a building.

SUMMARY

- The two most important concerns in installing a fiber-optic cable are minimum bend radius and maximum tensile load.
- Cables can be buried, strung aerially, or placed in trays and conduits.
- Closure/organizers are used to organize and protect spliced cable.
- Distribution boxes provide protection and organization for fibers and allow versatile and flexible distribution of fibers as application demands.
- Patch panels allow pluggable, rearrangeable interconnection of fibers.
- Wall outlets are similar to electrical outlets in that they allow equipment to be plugged into building wiring.

REVIEW QUESTIONS

1. What are the two most important factors to be considered when installing a cable?
2. Will a cable's minimum allowable bend radius be less during the installation or after?

3. Will a cable's maximum tensile load be greater during the installation or after?
4. Name three additional influences against which a buried cable must be protected.
5. In an intrabuilding application, what distinguishes an ordinary duplex cable from a plenum duplex cable?
6. When pulling a cable through a conduit, to what part must the pulling forces be applied?
7. Why is a fiber-optic rack either 19 in. or 23 in. wide?
8. What type of hardware allows connection between intrabuilding cable and drop cables from equipment?
9. What distinguishes a patch panel from a distribution box for splices?
10. If the total length of an application is twice the distance that can be achieved in a single pull, where is the best place to begin?

CHAPTER 15 Fiber-Optic Systems and Applications

This chapter looks at some typical applications using fiber optics. It gives special emphasis to local area networks, to the Fiber Distributed Data Interface, and to telecommunications. The chapter ends with a look at two emerging technologies that will affect future fiber-optic applications.

LOCAL AREA NETWORKS

A local area network (LAN) is an electronic communication network in which equipment in a limited geographical area, such as an office, a building, or a campus of buildings, is interconnected. All equipment is able to communicate with other equipment attached to the LAN. Attached equipment can be a personal computer, an engineering workstation, a minicomputer, and so forth; these are commonly referred to as stations or workstations. Each attachment point is a node. A network that has 50 attached workstations is a 50-node LAN. A node is an addressable point in the network, one that is capable of receiving and processing information. Occasionally, a node is a network device such as a bridge. A bridge allows different networks to be joined. To one network, the bridge appears as a single node, although this node actually represents another attached network. But the bridge remains capable of processing information, since it must pass the information to the attached network.

A LAN allows workers to share information databases, programs, files, and services such as electronic mail. Some applications distribute themselves among the different workstations on a LAN to allow different workers to work on the same information. For example, a document management system allows writers, illustrators, and editors all to work on the same document. While several illustrators are creating artwork or retouching photographs, writers can be writing different sections, editors can be editing, and designers can be working on the layout.

Since it is estimated that 80% of information used within a local environment is created in that same environment, a LAN provides an efficient way to allow users to share information and resources.

Most LANs operate at modest speeds over limited distances. Fiber optics is attractive for two reasons. First, it extends transmission distances. For example, the 10BASE-T network discussed below has a maximum distance of only 100 meters between hubs. By using a fiber-optic repeater at certain points in the network, the distance between hubs extends twenty times to 2 km to enlarge the geographical scope of the network.

The second reason is EMI immunity. A fiber-optic section is sometimes added to a LAN to combat noise.

In many applications, operating speed is not an issue dictating the use of fiber optics. Most copper-based LANs operate at either 10 or 16 Mbps, a quite modest speed for fiber optics.

LAN Topologies

A LAN is a sophisticated arrangement of hardware and software that allows stations to be interconnected and to pass information between them. The *topology* of a LAN refers to its physical and logical arrangement. Figure 15–1 shows the most common network topologies.

A bus structure has the transmission medium as a central bus from which each node is tapped. Messages flow in either direction on the bus.

The ring structure has each node connected serially with the node on either side. Messages flow from one node to the next in one direction around the ring.

The star topology has all nodes connected at a central point, through which all messages must pass.

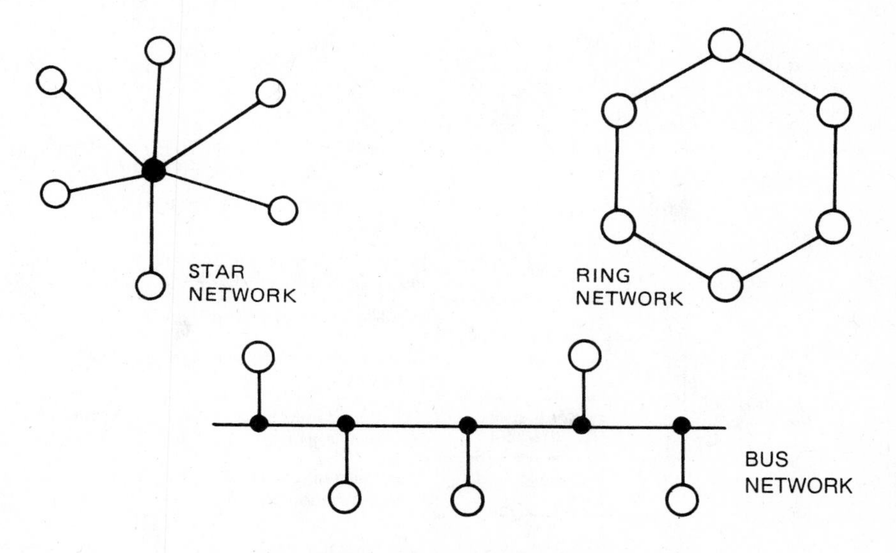

FIGURE 15–1 Network topologies

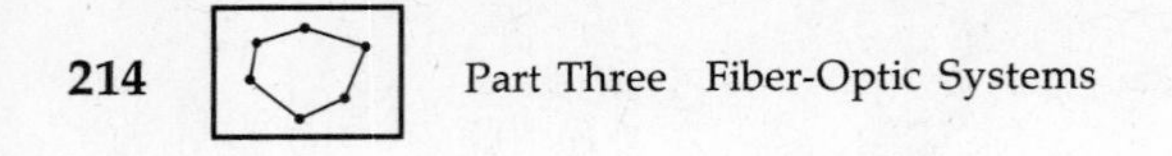

A hybrid topology combines more than one basic topology. For example, several stars can be connected in a ring or bus arrangement.

It is important to distinguish between a *physical* topology and a *logical* topology. A logical topology defines how the software thinks the network is structured—how the network is ''philosophically'' constructed. A physical topology defines how the LAN is physically built and interconnected. For example, the Ethernet LAN is a bus-structured network logically. Its physical topology depends on the specific variety of network. Sometimes it is wired like a traditional bus. Other times it is wired like a star or a hybrid configuration. Regardless of how it is wired or interconnected, the LAN appears to the control software as a bus network.

NETWORK LAYERS

As data communications becomes an important part of business, a need arises for universality in exchanging information within and between networks. In other words, there is a need for standardization, a structured approach to defining a network, its architecture, and the relationships between the functions.

The International Standards Organization (ISO), in 1978, issued a recommendation aimed at establishing greater conformity in the design of networks. The recommendation set forth the seven-layer model for a network architecture shown in Figure 15–2. This structure, known as the Open Systems Interconnection (OSI), provides a model for a common set of rules defining how parts of a network interact and exchange information. Each layer provides a specific set of services or

Internetworking Device		OSI Model Layer	Layer Function
	7	**Application**	Specialized functions, such as file transfer, terminal emulation or electronic mail
	6	**Presentation**	Data formatting and character code conversion
	5	**Session**	Negotiation and establishment of a connection with another node
↑Gateway	4	**Transport**	Provision for end-to-end delivery
Router	3	**Network**	Routing of information across multiple networks
Bridge	2	**Data Link**	Transfer of units of information, sequencing, and error checking
Repeater	1	**Physical**	Transmission of raw data over a communication channel

FIGURE 15–2 OSI seven-layer network model

functions to the overall network. The three lower layers involve data transmission and routing. The three top layers focus on user needs. The middle layer, transport, provides an interface between the lower layers and the top layers. Unfortunately, each layer has numerous standards that define the layer differently. The fiber-optic systems in this chapter, for example, define the lower layers differently.

The following is a brief description of each layer.

Physical Layer

The physical layer is the most basic, concerned with getting data from one point to another. This layer includes the basic electrical and mechanical aspects of interfacing to the transmission medium. This includes the cable, connectors, transmitters, receivers, and signaling techniques. Most of this book is concerned with the physical layer. Each specific type (or subtype) of network defines the physical layer differently.

While not used in LANs, the most widely used physical layer is RS-232, which defines the voltage levels of signals, the function of each wire, the type of connector, and so forth. The com or serial port of a personal computer implements an RS-232 physical layer. You can see that each specific need requires a different definition for this layer. Simply changing the cable type or connector creates a different layer.

Data-Link Layer

The data-link layer provides for reliable transfer of data across the physical link. This layer establishes the protocols or rules for transferring data across the physical layer. It puts strings of characters together into messages according to specific rules and manages access to and from the physical link. It ensures the proper sequence of transmitted data.

Network Layer

The network control layer addresses messages to determine their correct destination, routes messages, and controls the flow of messages between nodes.

Transport Layer

This layer provides end-to-end control once the transmission path is established. It allows exchange of information independent of the systems communicating or their location in the network.

Session Layer

The session layer controls system-dependent aspects of communication between specific nodes.

Presentation Layer

At the presentation layer, the effects of the layer begin to be apparent to the network user. This layer, for example, translates encoded data into a form for display on the computer screen or for printing. In other words, it formats data and converts characters. For example, most computers use the American Standard Code for Information Interchange (ASCII) format to represent characters. Some IBM equipment uses a different format, the Extended Binary Coded Decimal Information Code (EBCDIC). The presentation layer performs the translation between these two formats.

Application Layer

At the top of the OSI model is the application layer, which provides services directly to the user. Examples include resource sharing of printer or storage devices, network management, and file transfers. Electronic mail is a common application-layer program.

In practice, networks work from the top layer of one station (the message originator) to the top of another station (the message recipient). A message, such as electronic mail, is created in the top presentation layer of one workstation. The message works its way down through the layer until it is placed on the transmission medium by the physical layer. At the other end, the message is received by the physical layer and travels upward to the presentation level. It is at the presentation level that the electronic mail is read.

This layered approach to building a network holds two benefits. First, an open and standardized systems permits equipment from different vendors to work together. Second, it simplifies network design, especially when extending, enhancing, or modifying a network. For example, while most LANs use copper wire, either coaxial cable or twisted pairs, adding a fiber-optic point-to-point link involves only the physical layer. Higher levels of the OSI model do not care how the physical layer is actually implemented; they care only that the physical layer follows certain rules in interacting with higher levels.

A network can be connected to other networks of the same or different type. Sometimes users purposefully break a large network into several smaller segments to make the network more efficient.

Figure 15–2 also shows the devices required to interconnect either different networks or segments of the same network. The simplest device is a repeater operating at the physical layer. The repeater simply takes an attenuated signal, amplifies it, retimes it, and sends it on its way. It is a "dumb" device in that it does not look at the message; it only regenerates the pulses.

A bridge operates at the next level, the data link level, to link different networks that use the same protocol.

A router operates at the network-control level and can handle different protocols.

A gateway works at higher levels, serving as an entry point to a local area network from a larger information resource such as a mainframe computer or a telephone network.

ACCESS METHOD

Access refers to the method by which a station gains control of the network to send messages. Two popular methods are carrier sense, multiple access with collision detection (CSMA/CD) and token passing.

In CSMA/CD, each station has equal access to the network at any time (multiple access). Before seizing control and transmitting, a station first listens to the network to sense if another workstation is transmitting (carrier sense). If the station senses another message on the network, it does not gain access. It waits awhile and listens again for an idle network.

The possibility exists that two stations will listen and sense an idle network at the same time. Each will place its message on the network, where the messages will collide and become garbled. Therefore, collision detection is necessary. Once a collision is detected, the detecting station broadcasts a collision or jam signal to alert other stations that a collision has occurred. The stations will then wait a short period for the collision to clear and begin again.

In the token-passing network, a special message called a *token* is passed from node to node around the network. Only when it possesses the token is a node allowed to transmit.

FRAMES

The information on a network is organized in frames (also called packets). A frame includes not only the raw data, but a series of framing bytes necessary for transmission of the data. Figure 15–3 shows the frame formats for both the Ethernet and FDDI data transmissions. Other frames are also used. For example, besides the data frame, a token-ring network also uses a token frame for passing the token around the network.

Here is a brief description of the elements of an FDDI data frame.

The *preamble* indicates the beginning of a transmission. The preamble is a series of alternating 1's and 0's—1010101010 . . . —that allows the receiving stations to synchronize with the timing of the transmission.

The *start of frame* is a special signal pattern of 10101011 that signals the start of information.

The *packet ID* identifies the type of packet, such as data, token, and so forth.

The *destination address* is the address of one or more stations that are to receive a message. In a network, each station has a unique identifier known as its address.

Preamble	Start of Frame	Destination Address	Source Address	Length	Data	CRC
7 bytes	1 byte	6 bytes	6 bytes	2 bytes	46 - 1500 bytes	4 bytes

IEEE 802.3 Frame

Preamble	Start of Frame	Packet ID	Destination Address	Source Address	Data	CRC	End of Frame + Status
≥2 bytes	1 byte	1 byte	2 or 6 bytes	2 or 6 bytes	0 - 4486 bytes	4 bytes	≥2 bytes

FDDI Data Frame

FIGURE 15–3 Data frame formats

The *source address* identifies the station initiating the transmission.

The *data* is the information, the point of the transmission.

The *CRC* or *cyclic redundancy check* is a mathematical method for checking for errors in the transmission. When the source sends the data, it builds the CRC number based on the patterns of the data. The receiver does the same thing as its receives the data. The receiving station also builds a CRC. If the receiver CRC matches the transmitted CRC, no errors have occurred. If they don't match, an error is assumed, the transmitter is informed, and the transmission is sent again.

The *end of frame* informs the receiving station that the transmission is over. It also contains the status of the transmission. The receiving station marks the status to acknowledge receipt of the packet.

ETHERNET AND TOKEN RING

The two most popular LANs in use today are IEEE 802.3 Ethernet and 802.5 Token Ring. The numbers refer to the committees of the Institute of Electrical and Electronic Engineers that define the standards and to the documents approved as standards by the committee.

IEEE 802.5 Token Ring

IEEE 802.5 specifies a token-passing ring network. It is often called the IBM token ring because the original design was pioneered by IBM. While the LAN uses a *logical* ring, it is physically configured as a star network. Workstations attach through the network through a concentrator, and concentrators are connected by network cable called the backbone or trunk. The main application of fiber-optics in an IEEE 802.5 LAN is in interconnecting repeaters. The twisted-pair cable feeds the repeater, which converts the signal to optics and transmits it. The repeater unit at the other end receives the signal and converts it back to an electrical

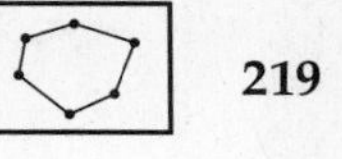

signal. Each repeater contains a port for the twisted-pair cable and transmit and receiver ports for the fiber-optic cable.

IEEE 802.3 Ethernet

IEEE 802.3 is a CSMA/CD LAN commonly called Ethernet. Ethernet, however, was originally a LAN defined by Xerox, Digital Equipment, and Intel that used a thick coaxial cable as the transmission medium. Since then, the standard has evolved considerably to be much more flexible, although Ethernet is still widely used to describe this network. Minor differences exist between the Ethernet specification and IEEE 802.3. Still, most people use the two terms interchangeably, and so shall we.

Table 15–1 lists several flavors of IEEE 802.3. The prefix number defines the operating speed, the middle word defines the signalling techniques, and the suffix defines the transmission medium. For example, 10BASE-T means a network operating at 10 Mbps, using baseband communication, and transmitting over twisted-pair cable.

While IEEE 802.3 is a logical bus topology, it is constructed as either a bus or a star depending on the type of cable used. Coaxial cables typically use a physical bus topology: The backbone cable runs serially from workstation to workstation. Twisted-pair and fiber use a logical star, with workstations connecting concentrator (also called a multiport repeater). Each concentrator contains several ports that permit attachment of a workstation or another concentrator. Figure 15–4 shows the difference between a pure bus and one using concentrators.

	Medium	Max. Segment Length (meters)	Nodes	Signaling Technique
10BASE5	Thick Coaxial	500	100	Baseband
10BASE2	Thin Coaxial	185	30	
10BASE-T	Twisted-Pair "Telephone" Cable	100	See Note	
10BASE-FP	Fiber-Optic Cable	1000	See Note	
10BASE-FB	Fiber-Optic Cable	2000	See Note	
10BASE-FL	Fiber-Optic Cable	2000	See Note	
10BROAD36	Coaxial Cable	3600	System dependent	Broadband

Note: Maximum number of nodes is limited either by timing constraints or by the number of repeater/concentrators in the network. Still, well over 100 nodes are possible.

Note: In late 1992, as this book was being published, a proposal was submitted to the 802.3 committee for "fast" Ethernet or 10BASE-VG. This would provide 100 Mbps transmission speed over twisted-pair cable.

TABLE 15–1 IEEE 802.3 network varieties

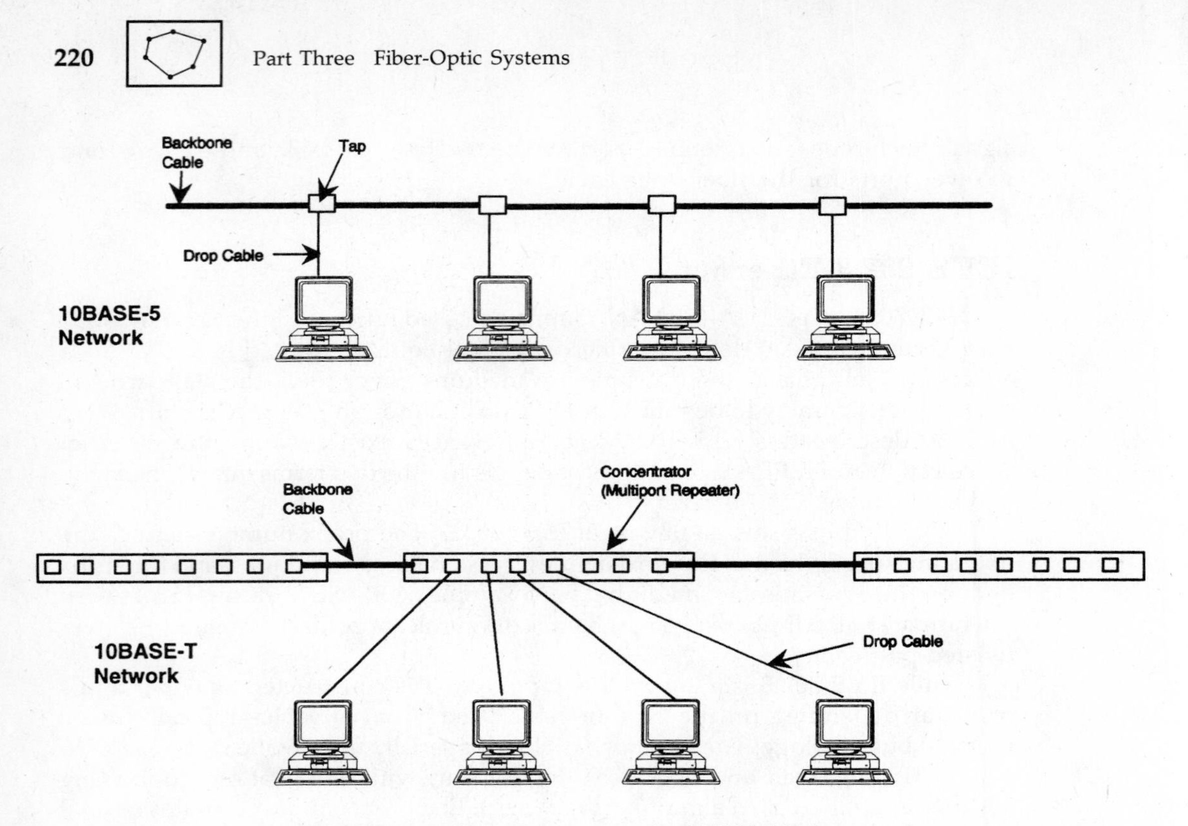

FIGURE 15–4 Bus versus star-wired IEEE 802.3 networks

A standard port acts as a transceiver for the type of cable being used; one function of the transceiver is to ensure that the signal is converted to the form required by the cable.

For example, a twisted-pair cable uses four wires, two for transmitting and two for receiving, and a modular-jack telephone connector. A coaxial cable uses only a single cable, allowing transmission in both directions over a single wire, and either BNC or N-type connectors. A fiber-optic cable requires two fibers, one for transmitting in each direction, and usually ST or SMA connectors.

Another type of port is the attachment unit interface (AUI). The AUI port is a 15-pin subminiature-D connector (similar to the 25-pin connector used in serial ports on personal computers) that does not include the transceiver function.

An AUI allows different transceivers to be used. For example, consider a network adapter board in a workstation. You can buy a board specific to the type of cable you are using: a 10BASE-T board for twisted-pair cable or a 10BASE-2 board for thin coaxial cable. Alternatively, you can buy a board with a AUI port. The AUI port allows you to plug in a transceiver for whatever cable you wish to use: twisted pair, thick or thin coaxial, or fiber optic. The FOIRL fiber-optic transceiver discussed below always plugs into an AUI port.

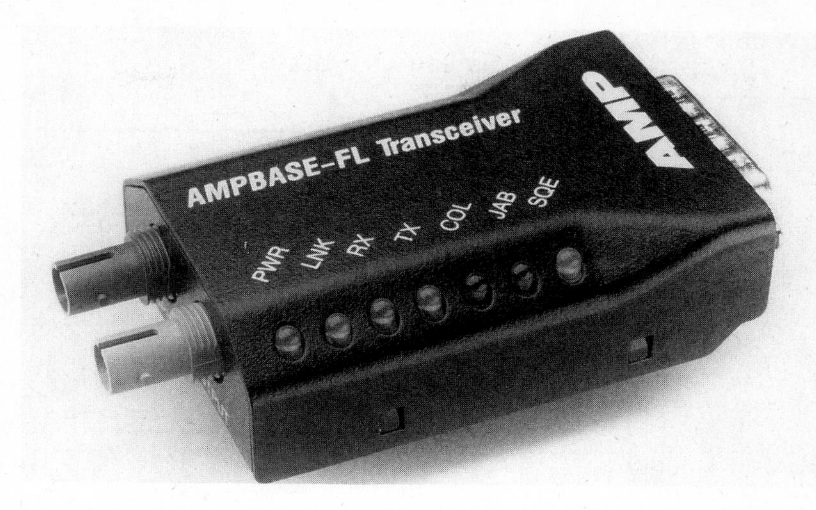

FIGURE 15–5 FOIRL/10BASE-FL10 transceiver (Photo courtesy of AMP Incorporated)

FOIRL

We have discussed how a repeater accepts attenuated data, amplifies and reshapes it, and sends it on its way. A fiber-optic interrepeater link (FOIRL) uses a fiber-optic transceiver at each end. The FOIRL transceiver (an example suited to both FOIRL and 10BASE-FL applications is shown in Figure 15-5) is strictly intended to connect two repeaters, but versions are available that also allow attachment to a workstation. A common use of a FOIRL unit is to achieve extended distances or provide additional EMI immunity in a noisy environment. While the FOIRL specification specifies a transmission distance of 2 km, it assumes a 62.5/125-μm fiber having an attenuation of 3.75 dB/km. A lower loss cable can extend transmission distances to over 3 km. But fiber does not allow the distance to be extended indefinitely. The collision-detection protocol ultimately limits the distance because of timing restraints in the network.

Figure 15–6 shows a typical application in a 10BASE-T network. This network uses twisted-pair cable; all workstations are connected to one another through concentrators, and concentrators are interconnected to increase the size of the network. Twisted-pair cable has a limit of 100 meters between concentrators or between concentrator and workstation. By using a FOIRL transceiver on the AUI port, distances are greatly extended. For example, LAN concentrators are typically placed in a wiring closet. The distance between closets is determined by the

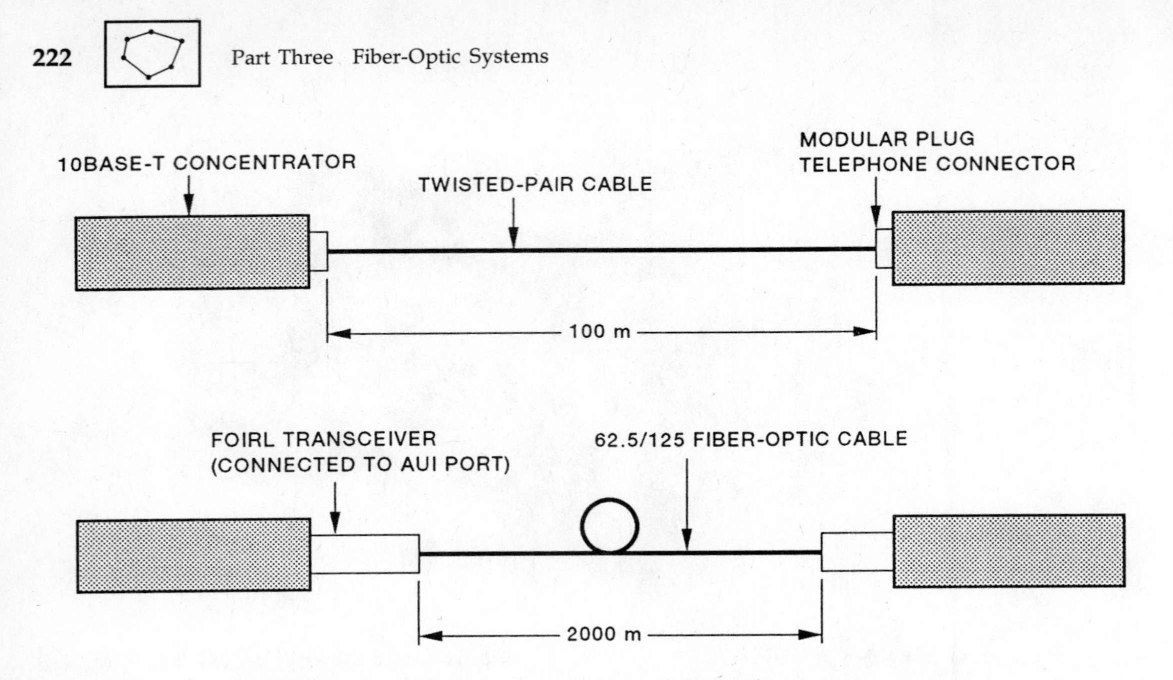

FIGURE 15–6 FOIRL extends distances in copper-based networks

allowed cable distance. A fiber-optic link allows concentrators not only to be easily connected between floors of a building, but between buildings as well.

10BASE-F

The 10BASE-F specification adds a fiber-optic alternative for use in IEEE 802.3 networks. The specification actually lists three different variations: a backbone cable (10BASE-FB), a star-coupled network (10BASE-FP), and a fiber-optic link between concentrator and station (10BASE-FL). Figure 15–7 shows a typical network containing all three variations; while Table 15–2 is a table summarizing some of the principal optical characteristics of the 10BASE-F network.

The requirements for a 10BASE-FB and a 10BASE-FL are fairly straightforward. The backbone is simply a point-to-point connection between concentrators. The FL versions define connection between a concentrator and station or between concentrator and passive star coupler. Notice that the concentrator is a multiport repeater that contains electronics to regenerate the signal. Each point-to-point link can be up to 2 km long.

The 10BASE-FP option uses a passive fiber-optic star coupler to distribute light. The star can contain up to 33 ports. Ports can attach to a workstation or to a 10BASE-FL concentrator. The distance from the star to the attached station or concentrator is 500 m. Because the star coupler is passive, a 10BASE-FP network can connect workstations over an all-fiber network, with no electronics involved outside the workstation.

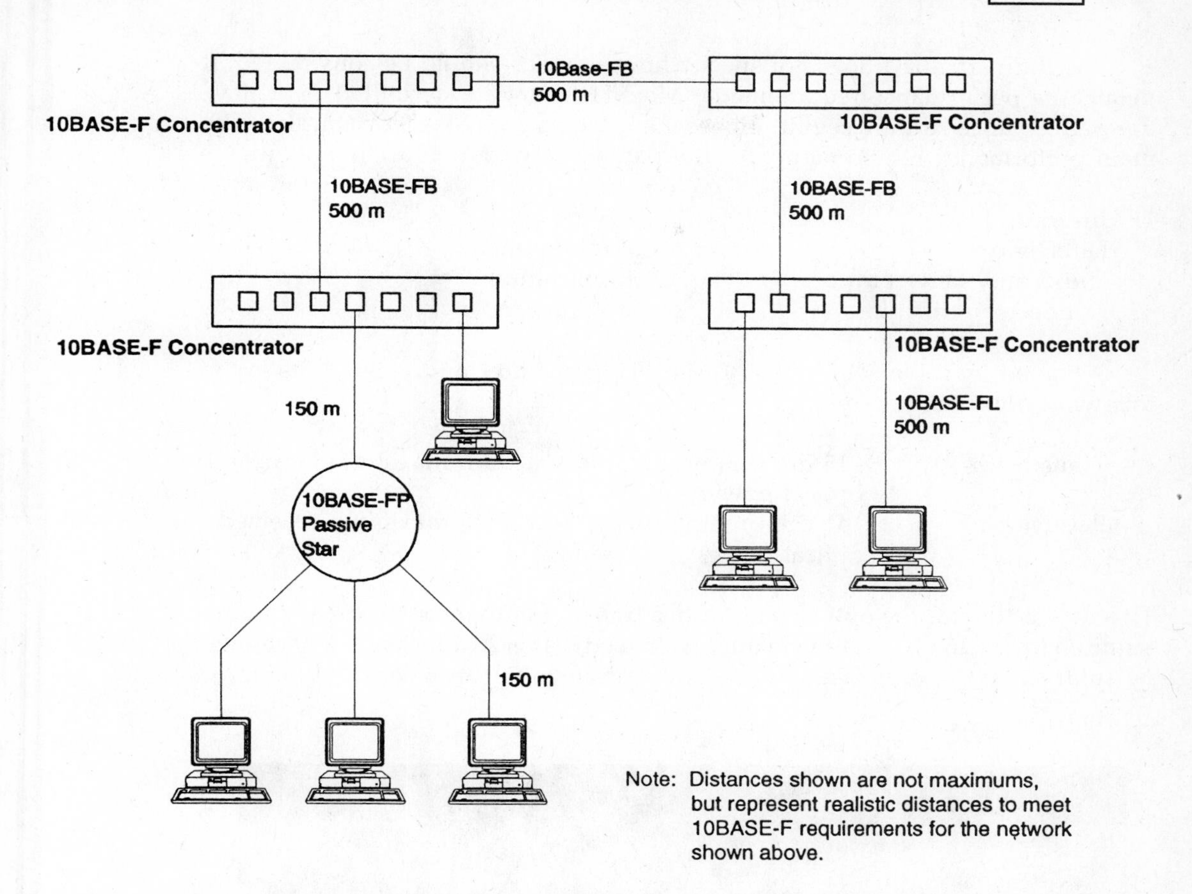

FIGURE 15–7 Typical 10BASE-F network

	10BASE-FP		10BASE-FB		10BASE-FL	
	Min	**Max**	**Min**	**Max**	**Min**	**Max**
Optical Wavelength (nm)	800	910	800	910	800	910
Spectral Width (nm)		75		75		75
Risetime (ns)	2	10		10		10
Launch Power (dBm)	–15	–11	–20	–12	–20	–12
Received Power (dBm)	–41	–27	-32.5	–12	-32.5	–12
Max Length (m)		500		2000		2000
Connector Type	ST		ST		ST	

TABLE 15–2 10BASE-F optical performance requirements

The IEEE standard does not specify how a device should be constructed; it details the performance requirements. Manufacturers are permitted to achieve the performance as they see fit. Figure 15–8 shows a passive star coupler. The main performance requirements for the passive star coupler are as follows:

Insertion loss:	16 to 20 dB
Directivity:	35 dB minimum
Uniformity between ports:	2.5 dB maximum
Connector return loss:	25 dB minimum

As listed in Table 15–2, the 10BASE-FP transmitter and receiver have the following power requirements:

Transmitter:	–15 dBm minimum and –11 dBm maximum launched optical power
Receiver:	–41 dBm minimum and –27 dBm maximum received optical power

This means that for the optical path from a transmitter to receiver, the minimum required loss is 16 dB and the maximum allowed loss is 26 dB. This is determined by subtracting the best-case and worst-case transmitter/receiver combinations:

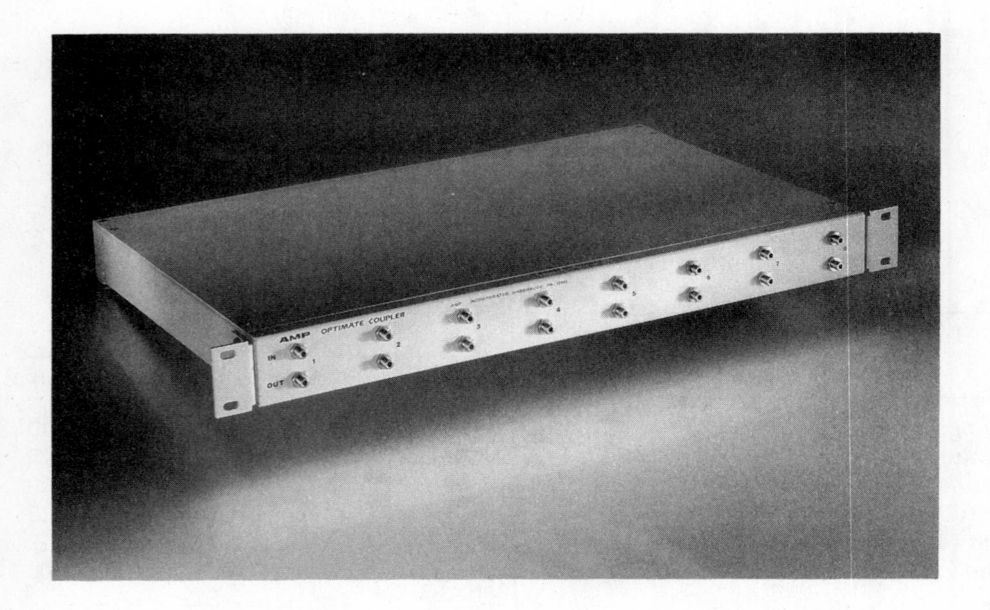

FIGURE 15–8 10BASE-FP star coupler (Photo courtesy of AMP Incorporated)

–11 dBm – (–27) dBm = 16 dB
–15 dBm – (–41) dBm = 26 dB

In other words, if the transmitter launches a full –11 dBm into the fiber, at least 16 dB must be lost over the transmission path. The receiver is specified to receive no more than –27 dBm. If, for example, the path contributes only 10 dB of loss, power at the receiver will be –21 dBm, which is more than allowed. The standard recognizes that in some cases it may be necessary to add loss to satisfy the received-power requirement.

Similarly, if the transmitter launches the worst-case power of –15 dBm and a full 26 dB of loss occurs in the transmission path, the received power is still –41 dBm, which falls within the limits of the specification.

The minimum loss is that contributed only by the star coupler. The maximum loss includes the coupler, cable, and any intervening connectors at a wall outlet or path panel.

FIBER DISTRIBUTED DATA INTERFACE (FDDI)

The Fiber Distributed Data Interface—FDDI—is the first local area network designed from the ground up to use fiber optics. Compared to its copper-based counterparts, its performance is quite impressive: a 100-Mbps data rate over a 100-km distance and having up to 1000 attached stations. Compare this to the 16-Mbps rate for an 802.5 token ring or 10-Mbps rate for 802.3 Ethernet and interconnection distances measured in hundreds of meters, and you'll see that FDDI offers considerable performance advantages.

Some of the pertinent optical characteristics of FDDI are as follows:

Source:	LED or laser
Operating wavelength:	1300 nm
Power launched into fiber:	–20 dBm minimum
Fiber type:	62.5/125-μm multimode
	Single mode
Maximum nodes:	1000
Maximum total length:	100 km
Distance between stations:	2 km, multimode
	40 km, single mode
Power budget between nodes:	11 dB

While the FDDI standard recommends 62.5/125-μm fiber for multimode applications, it also allows 50/125-, 85/125-, and 100/140-μm fibers as long as these do not exceed the power budget or distort the signal beyond the limits allowed by the specification. The power budget does not include optical loss at the interface between the source and connector. The specification calls for a minimum power to be launched into the fiber. This simplifies loss calculations. The power budget between stations must only consider the fiber and any interconnections along the path.

FDDI meets the growing need for greater bandwidth in LANs. One trend in computing in the 1900s is termed ''downsizing.'' Large mainframe computers are being replaced by LANs using personal computers and workstations to distribute the work and the information. While slower speed LANs will predominate, their information-carrying capacity can be quickly taxed. FDDI provides much greater performances (often called throughput in computer jargon).

FDDI Topology

FDDI is arranged as a token-passing ring topology that uses two counter-rotating rings as shown in Figure 15–9. The primary ring carries information around the ring in one direction, while the secondary ring carries information in the other direction. The reason for two rings is redundancy: If one ring fails, the other is still available. Redundancy lessens the likelihood of network failure. Further protection lies in the fact that each station is attached only to adjacent stations in the ring. If a station fails or a single point-to-point link fails, the network still functions. If a cable break occurs between two stations, a station can accept data on the primary link and transmit on the secondary link, as shown in Figure 15–10. This wraparound function makes FDDI highly reliable.

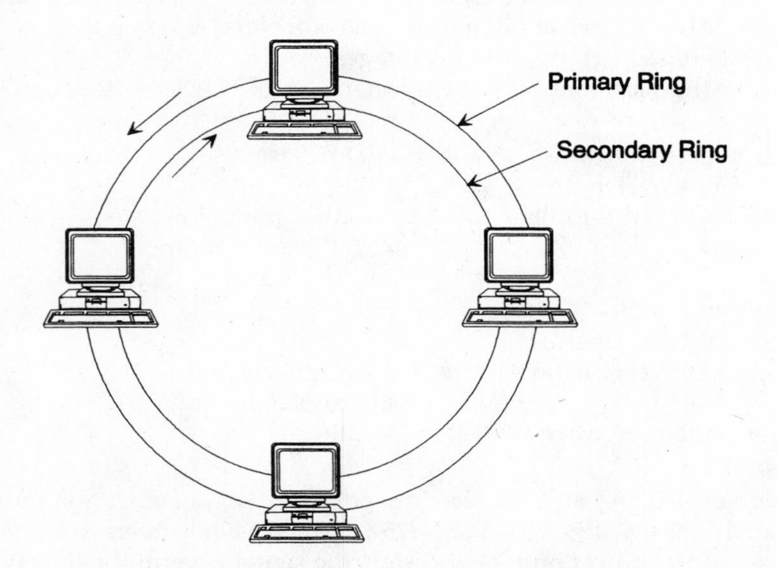

FIGURE 15–9 FDDI network

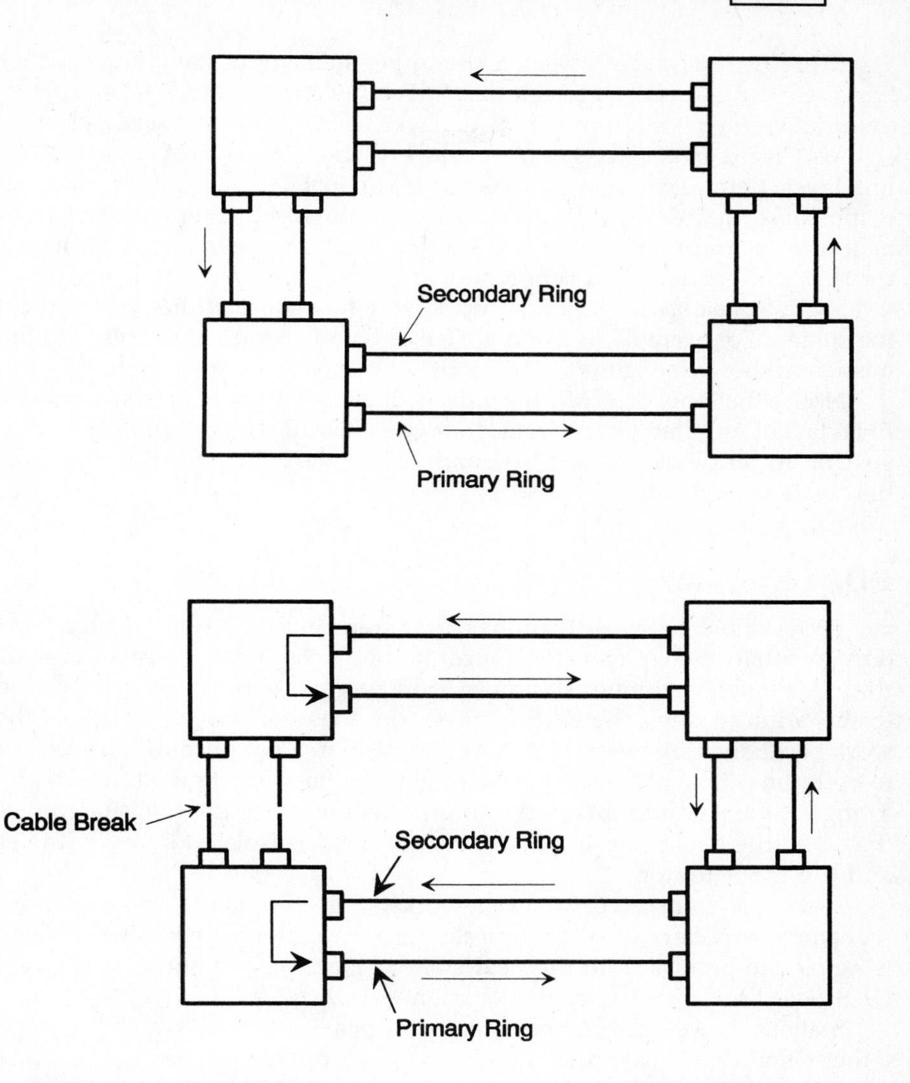

FIGURE 15–10 FDDI fault tolerance with primary and secondary rings

When the secondary ring is not needed for redundant operation, both the primary and secondary rings can be used for data transmission to effectively double the network transmission speed to 200 Mbps.

The FDDI network is defined by four documents:

PMD: The *physical medium dependent* is FDDI's lowest sublayer and corresponds to the OSI physical layer. As its title suggests, it defines the requirements for the physical transmission medium: transceiver, optical power, media interface connector (MIC), and cable. This is the layer that deals with fiber optics.

PHY: The *physical layer* also is the upper sublayer of the OSI physical layer. Serving as an interface between the PMD layer and the MAC layer, its function is to deliver symbols from the MAC layer to the FDDI physical network.

MAC: The *media access control,* which is the lower sublevel of the OSI data link layer, defines the frame formats used by FDDI (see Figure 15–3). Since it is responsible for controlling the flow of data, the MAC layer checks each packet as it arrives, transfers the data to the workstation's memory if the message is for that workstation, and sends the packat on its way.

SMT: *Station management* is a supervisory function that monitors and controls the station. For example, it is the SMT that detects a fault and redirects the transmission as shown in Figure 15-10. The SMT function is not part of the OSI hierarchy.

Notice that only the PMD layer deals directly with fiber optics. The MAC and PHY layers are electronic. Semiconductor manufacturers offer chips sets that specifically implement the MAC and PHY layers. The SMT function can be hardware or software.

FDDI Stations

Each station contains either one or two transceivers. A dual attachment station (DAS) contains two transceivers for connection to both the primary and secondary ring. A single attachment station (SAS) has a single transceiver for connection to the primary ring. The SAS workstation does not attach directly to the network backbone but goes through a concentrator. The concentrator may connect to both the primary and secondary ring of the fiber backbone, but only through a single connection to the workstation. This preserves the fault-tolerant redundancy on the backbone. If a SAS fails, the fault is isolated between the station and the concentrator.

Having a transceiver in each workstation eliminates the need for taps, repeaters, amplifiers, or other external signal conditioning equipment. Each cable is a point-to-point link to the next station. Figure 15–11 shows a typical FDDI DAS board for a computer workstation.

Stations, however, are often equipped with a dual bypass switch (Figure 15–12). If the station fails (or is turned off), the station is optically bypassed by connecting the input/output lines of the fiber backbone cable.

Since each station or concentrator is a repeater, you might think that the network could be extended to an infinite number of nodes over an infinite area. After all, the signal is regenerated at each station. FDDI, however, places another limit on the network: the round-trip time of a message from the transmitting station. When a station transmits, the message goes from station to station. Each station reads the message, checking to see if it is the recipient and checking for errors. It then passes the message to the next station. When the destination station receives the message, it changes the status byte on the frame. The station then continues around the ring back to the transmitting station. The station verifies that the message has been received by checking the status byte and that no errors

FIGURE 15–11 FDDI station board (Photo courtesy of Network Peripherals Inc.)

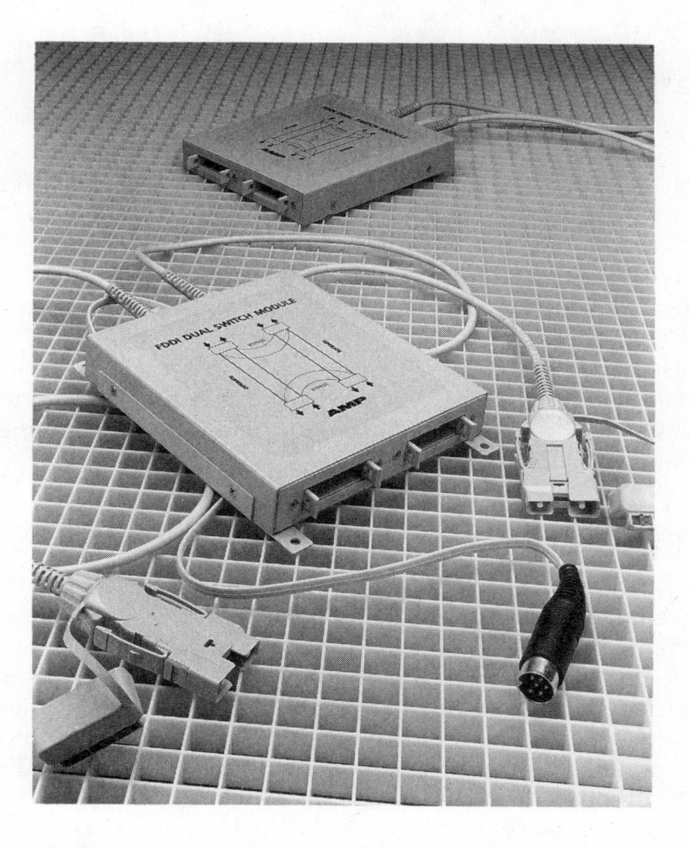

FIGURE 15–12 FDDI dual bypass switch for both primary and secondary rings (Courtesy of AMP Incorporated)

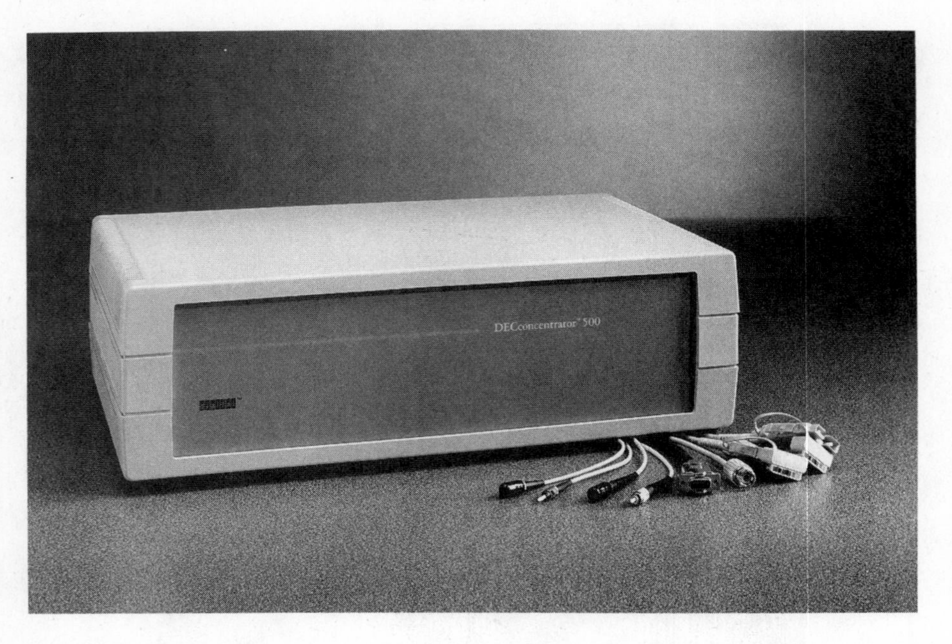

FIGURE 15–13 FDDI concentrator (Courtesy Digital Equipment Corporation)

have occurred in the message's content. The maximum number of nodes and network length is limited by the time it takes this round trip. This time is adjustable in the SMT software to allow very large networks to exist.

FDDI also uses concentrators (Figure 15–13) to connect to both the primary and secondary backbone rings, but only from a single connection to SAS workstations. This simplifies the network layout and cuts costs without endangering the redundancy of a dual backbone.

FDDI Applications

An FDDI network is often used in conjunction with other networks. It can be used as a backend network, a frontend network, or a backbone network as shown in Figure 15–14.

In a backend network, FDDI is used to connect mainframe computers, storage devices, controllers, and peripherals. Communication between these devices is one of the bottlenecks that slows computer operation. FDDI keeps the data zipping along. This application is more commonly done by the ESCON and Fiber Channel architectures discussed later in the chapter.

As a frontend network, FDDI connects to high-end workstations used in such applications as computer-aided design, simulation and control, and high-end

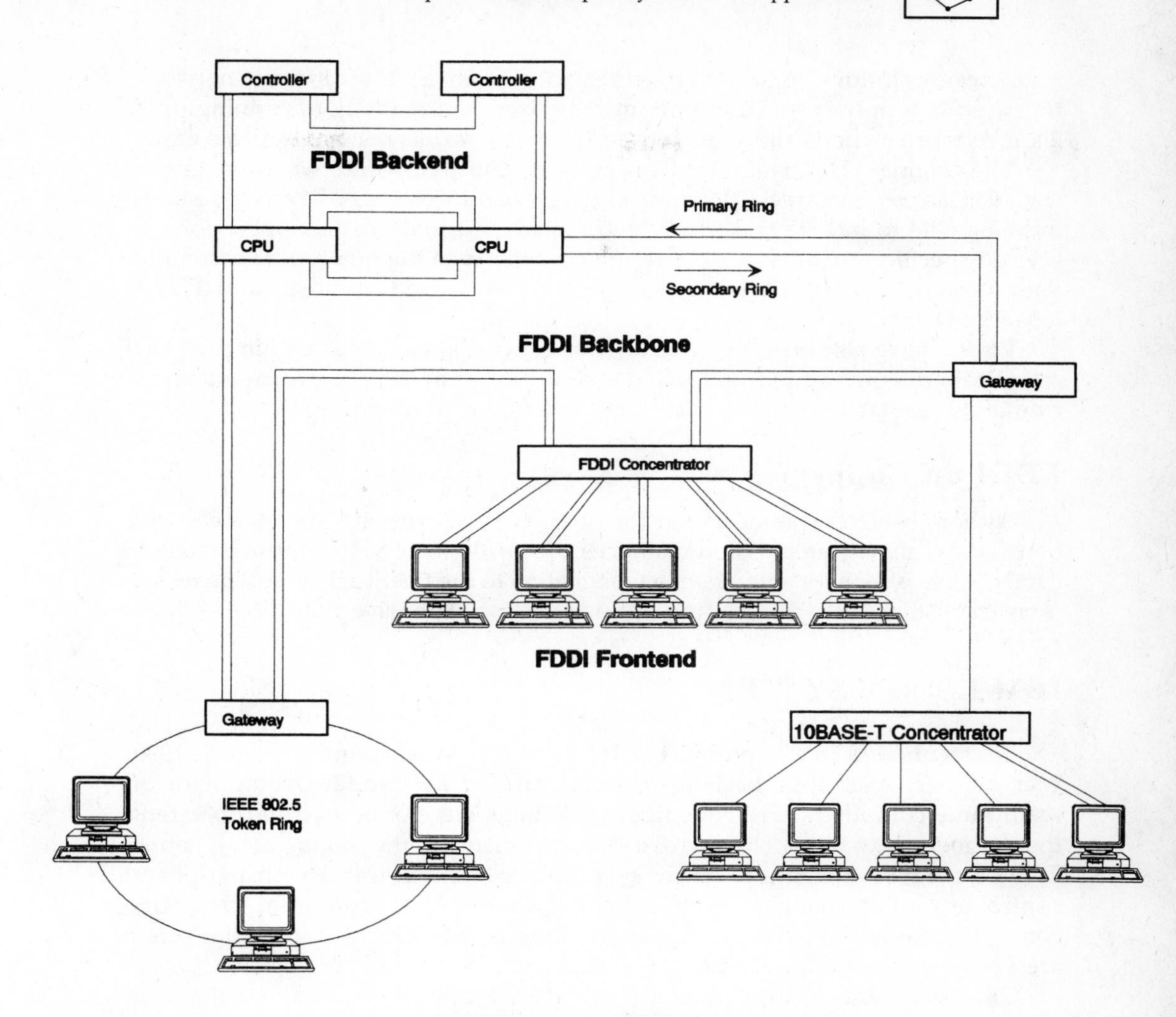

FIGURE 15–14 FDDI network

publishing. Such applications usually generate large files that can change often. Files, for example, can run to several million bytes. A color photograph can require a file of a couple of million bytes; in contrast the entire file for the text of this book is only about 750,000 bytes. FDDI has the data-transmission speed to prevent (or lessen) bottlenecks in transmitting information to and from workstations.

Finally, FDDI can serve as a backbone network connecting other networks. For example, a large company could have an 802.3 or 802.5 network for each of its various departments—engineering, accounting, sales, manufacturing, marketing, and so forth. Each individual network connects to the FDDI backbone through a gateway or bridge. FDDI then serves to connect all the different networks together. Not only is speed important here, but the capability of running up to 2 km between stations allows widely separated networks to communicate easily.

The original FDDI is an asynchronous system that works well transferring such data as computer files. However, the wide bandwidth of FDDI creates interest in being able to handle real-time video, voice, and multimedia applications. A revised specification adds a synchronous capability to the network to accommodate these needs. This revision is properly called hybrid ring control (HRC) or loosely FDDI II.

Efforts have also been made to offer lower cost alternatives within the FDDI specification. For example, simplified methods of connecting workstations to the ring allow use of lower-powered LEDs and SC or ST connectors.

FDDI on Copper

Although FDDI was designed for fiber, twisted-pair copper cable can also be used for short transmission distances up to 100 meters. The primary use of copper cable is for short runs from a workstation to the fiber backbone. The reason is mainly economics: The copper link is still less expensive than fiber today.

IBM ESCON SYSTEM

As mentioned in Chapter 1, the IBM ESCON system, introduced in 1990, was the first fiber-optic backend channel offered as standard equipment on mainframe computers. Previous fiber-optic links were options, used to extend the channel length but not to improve the performance of the channel. The channel is the means by which a mainframe computer communicates with peripheral control units. For example, you have seen the banks of tape drives in some large computer sites. While the tapes store enormous amounts of information, they are very slow in comparison to the host mainframe computer.

The peripheral control units are specialized computers that manage such peripherals as tape drives and large disk drives. The mainframe requests the information from the peripheral controller; the controller locates it on the drives and sends it to the mainframe in a fast burst. But even this burst is slow compared to the operating speed of the mainframe. Thus, whenever the mainframe must send or retrieve information over the channel, it slows down. The channel also has a limited distance.

The standard copper-based IBM channel operates at 4.5 Mbps over a maximum of 400 feet. Signal transmission is in parallel, meaning each bit of a byte travels on a separate wire.

The first version of the ECON system operates at 10 Mbyte/s, although a 17 Mbyte/s option is also available. (Notice that we said *bytes,* not *bits.*) But because operation is serial over a single fiber, the transmission speed is 200 Mbps. The system uses the 8B/10B code discussed in Chapter 7.

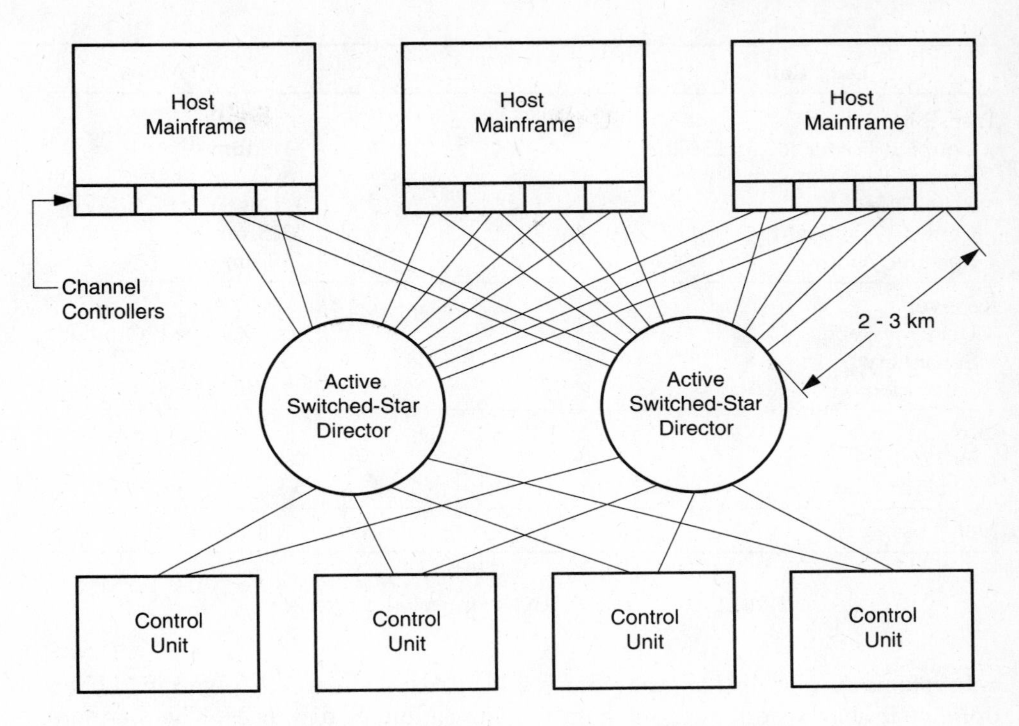

FIGURE 15–15 Example of IBM ESCON system

As shown in Figure 15–15, each mainframe can have several channels. In a copper-based system, each channel connects to a single control unit. The ESCON system uses an active switched-star configuration. The star, which is called a director by IBM, is an electronic subsystem that routes requests from the mainframe to the proper control unit. Unlike a passive star coupler, it provides output only on lines needed, not on all lines. It also allows several mainframes to share the same control unit. A channel may encounter up to two Directors in series. Each Director contains from 8 to 60 ports. Because each host computer can have several channel controllers, the system topology can become quite complex.

The ESCON system uses 50/125- or 62.5/125-μm fibers operating at 1300 nm. Each link segment can be 2 or 3 km (the 3 km is for low-loss 62.5/125-μm fiber). A mainframe-to-controller interconnection can pass through two Directors to achieve distances of 9 km. An optical laser system can extend distances up to 60 km.

Table 15–3 summarizes the optical characteristics of the ESCON system.

FIBER CHANNEL

Fiber Channel serves the same purpose as the ESCON system: It is a high-speed channel interconnecting a computer mainframe and various peripheral

Data Rate	200 Mbps
Transmitter	
Coupled Power (62.5/125 Fiber)	–17 dBm typ
Source Type	InGaAsP LED
Rise Time	1.2 ns
Center Wavelength	1325 nm
Spectral Width	150 nm
Receiver	
Detector Type	InGaAs Pin Photodiode
Sensitivity	–35 dBm avg
Saturation	–13 dBm
Link Distance	
50/125 Fiber	2 km
62.5/125 Fiber	3 km
Link Loss	8 dB max

TABLE 15–3 IBM ESCON optical characteristics

subsystems such as disk or tape storage. Fiber Channel is an ANSI-specified fiber-optic channel that offers faster data rates over longer distances. The standard allows several optical levels to meet cost and performance goals of various applications. Table 15–4 summarizes the levels. Notice the use of a short-wavelength laser operating at 780 nm. The laser allows a fast operating speed but considerably less distance than the long-wavelength laser. Very similar to the lasers used in compact-disk players, the 780-nm laser is less expensive than a 1300-nm laser. Fiber Channel uses the same 8/10B encoding scheme as the ESCON system.

			Performance		
Application	**Transmitter**	**Fiber Type**	**Mbaud**	**Mbyte/s**	**Distance (km)**
Low End	1300-nm LED	62.5/125 Multimode	132.81	5	1
Midrange	1300-nm Laser	Single Mode	265.65	25	30
	1300-nm LED	62.5/125 Multimode	265.65	25	1
	780-nm Laser	50/125 Multimode	265.65	25	2
High End	780-nm Laser	50/125 Multimode	531.25	50	1
	1300-nm Laser	Single Mode	531.25	50	10
	1300-nm Laser	Single Mode	1062.5	100	2, 10

TABLE 15–4 Fiber channel

TELECOMMUNICATIONS

The telecommunication industry is the heaviest user of fiber optics because of its high bandwidth, loss losses, and electrical immunity. The two most commonly used fibers are 62.5/125-μm multimode and single mode.

The earliest uses of fiber optics were trunk lines connecting central offices and long-distance toll centers. The central office contains switching equipment providing telephone service over a given geographical area. The central office is designated by the first three numbers of a seven-digit telephone office. Toll centers contain switching equipment for distributing calls from one metropolitan area to another.

Fiber is also finding application in the subscriber loop, which is the circuit connecting a central office to subscriber telephones. From the central office, feeder lines connect to a smaller electronic system called the *serving area interface*. From this interface go the lines that connect directly into the telephones in the home. The bandwidth of fiber may allow the telephone company to provide additional services to the home, including video, information services, and other bandwidth-hungry applications. (Current law prevents telephone companies from delivering video services.) Figure 15–16 shows a simplified diagram of a telephone network.

The 1990s are the decade when fiber will be extensively applied in the subscriber loop. Two approaches are used: fiber to the curb (FTTC) and fiber

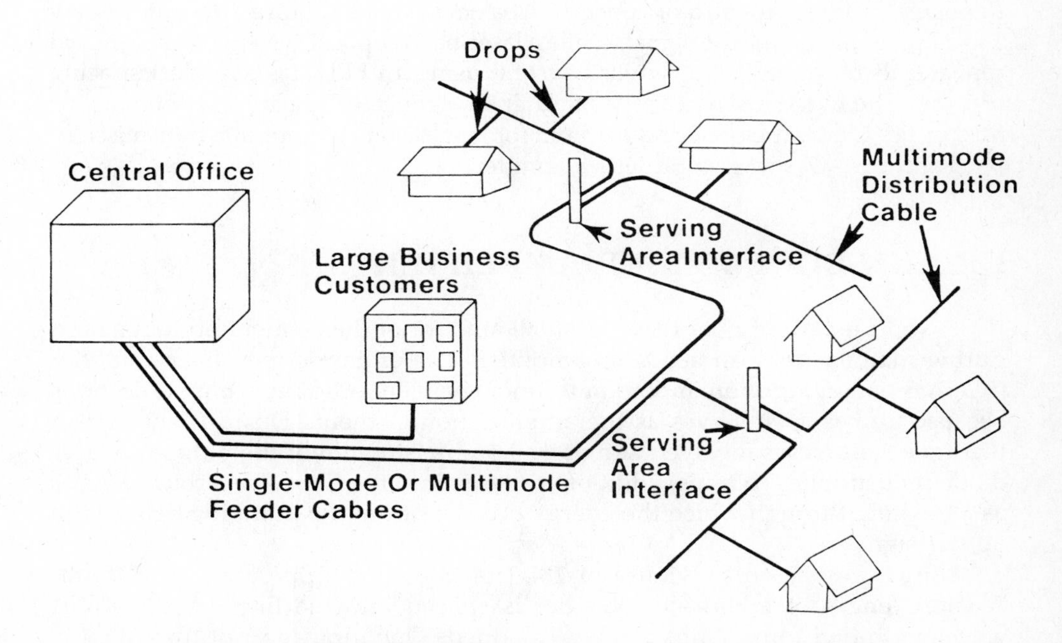

FIGURE 15–16 Telephone system (Courtesy of Corning Glass Works)

to the home (FTTH). In FTTH, the fiber transceiver is inside the home. Distribution inside the house can be by fiber or copper. In FTTC, the transceiver remains outside the home on the subscriber circuit. Copper drop cables carry signals to and from several homes. Because it is more economical, this approach is expected to find widespread application.

SONET

As mentioned in Chapter 3, Sonet (synchronous optical network) is a fiber-optic telecommunications standard providing data transmission rates up to 10 Gbps.

Sonet was first proposed in 1984 as a standardized method of transmitting optical signals and for ensuring that equipment from different companies is able to work together. Major telephone companies and telecommunications-equipment suppliers adopted Sonet in 1988. The basic Sonet rate is 51.840 Mbps, termed Optical Carrier 1 or OC-1. Higher speeds are simply multiples of the basic rate: OC-n = n × 51.840. For example, OC-3 has a rate of 155.52 (51.840 × 3 = 155.52). The highest rate, currently OC-192 operating at approximately 10 Gbps, is limited by the commercial availability of electronic components operating at higher speeds. As these components become available, Sonet standards will be expanded to include higher data rates.

Notice that Sonet transmission rates begin where coaxial cable rates end. Repeater spacings are also of concern. Coaxial systems require a repeater every 1 to 2 km to regenerate the signal, while a laser-based optical system has a standard repeater distance of 25 km, while a system using an LED has a modest spacing of 2 km. The much higher capacity and greater repeater spacing are the reasons fiber optics is now the preferred medium for long-distance telephone transmission.

Figure 15–17 shows a Sonet transmitter.

ERBIUM-DOPED FIBER AMPLIFIER

A short length of fiber that has small amount of the element erbium added during manufacture can act as an amplifier. The erbium-doped fiber amplifier (EDFA) is emerging as an important technology in long-distance communications because a fiber itself serves as the amplification element. This does not mean that no electronics is involved. An optical pump is required; the pump is a laser diode that supplies (pumps) additional energy into the fiber. The erbium serves as a medium through which the energy transfers from the pumping light to the optical signal.

Here is how it works (Figure 15–18). Consider a fiber operating at 1550 nm. A short length of erbium-doped fiber is spliced into the fiber. A three-port wavelength-division multiplexer is also added. One input port of the WDM is

FIGURE 15–17 Sonet transmission equipment (Courtesy of Akatel)

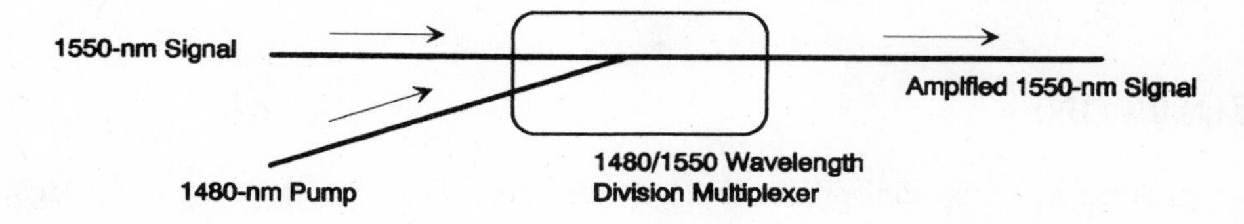

FIGURE 15–18 Erbium-doped fiber amplifier

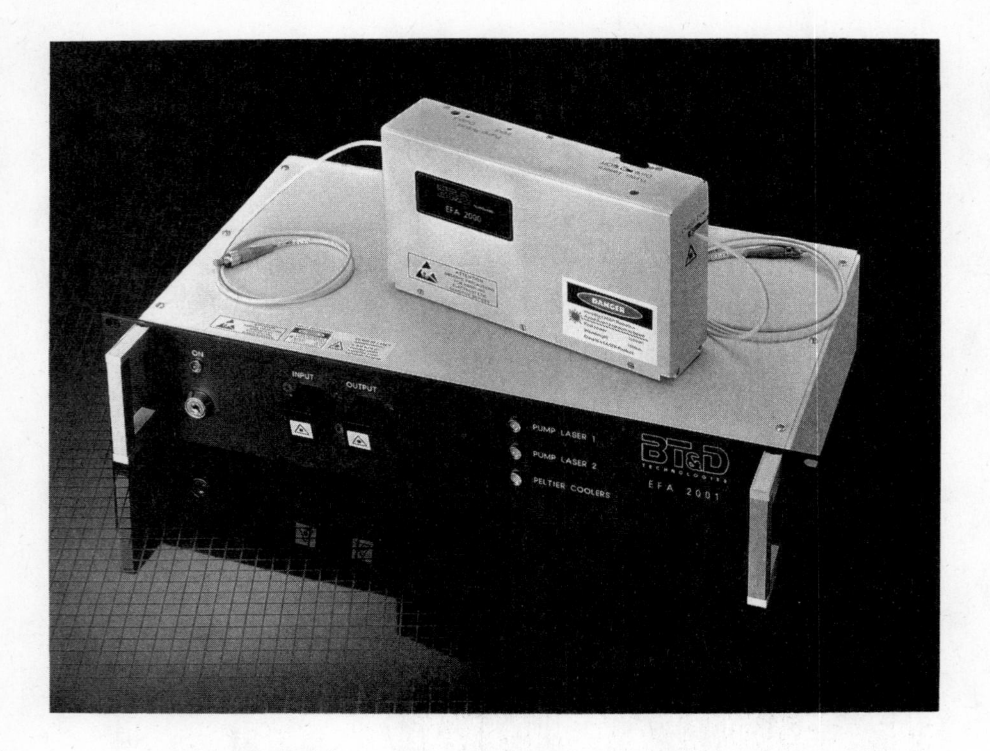

FIGURE 15–19 Example of EDFA (Courtesy of BT&D)

the 1550-nm signal fiber. The other input port is attached to a laser operating at either 980 nm or 1480 nm. The pumping light energizes the erbium. When the energized erbium loses it extra energy, it transfers it to the 1550-nm signal, amplifying it up to 30 dB. Because the pumping wavelength and the signal wavelength are different, they do not interfere with each other. Experiments with EDFAs have boosted transmission distances by a factor of 100.

Figure 15–19 shows an EDFA. Adding the erbium-doped fiber and WDM is fairly simple, and the electronics for the laser pump is simpler than for an electronic-based repeater. An important feature that will lead to widespread use of EDFAs is that amplification takes place without regard to signal rate or other characteristics.

SOLITONS

A soliton is a special pulse of light that does not disperse as it propagates. A soliton maintains its shape indefinitely in a perfect fiber, neither broadening nor compressing. In actual fibers, the pulse will be attenuated. If amplification is added, perhaps by an EDFA, so that pulse amplitude remains constant, soliton behavior can be generated in practice. While the exact physics of solitons is complex, the essential effect is to create two countering effects. Dispersion, you recall, occurs because different wavelengths travel at different velocities. Another phenomenon—called the optical Kerr effect—is that above a certain level, different intensities of the *same* wavelength also travel at different speeds. A soliton uses the optical Kerr effect to counteract dispersion. In other words, while dispersion tries to broaden the pulse, the optical Kerr effect tries to compress it. When the two phenomena are balanced, the pulse stays exactly the same shape.

Linn Mollenauer, a researcher at Bell Labs who pioneered soliton research, likens the soliton to three runners on a mattress. The center runner is in a trough formed by the weight of the runners. The front runner (representing dispersion) is slowed because he is running up the slope from the trough. The rear runner (representing the optical Kerr effect) is pulled along faster as he moves down the hill into the trough. As the runners move, the front runner never gains and the rear runner never falls behind. They stay perfectly together.

In 1988, Mollenauer demonstrated solitons by injecting solitons into a 40-km reel of fiber and then closing the loop. The solitons traveled round and round through the fiber for over 10,000 km with little broadening.

Whereas EDFAs are commercially available, soliton systems have not been used in real-world application as of late 1992. But the intense interest in solitons means they will find application, moving from the laboratory to the real world soon. The combination of solitons and EDFAs promises fiber-optical transmission systems of lengths far surpassing those of today's systems. Today, we speak mainly of megabit systems over tens of kilometers. Solitons and EDFAs promise gigabit systems over thousands of kilometers.

SUMMARY

- Many fiber-optic applications have been standardized.
- Local area networks use fiber optics.
- Fiber optics finds use in copper-based LANs to extend transmission distances.
- IEEE 802.5 is a token-passing ring network operating at 4 or 16 Mbps.
- IEEE 802.3 is a CSMA/CD bus-based LAN operating at 10 Mbps and using coaxial cable, twisted-pair cable, and fiber-optic cable.
- 10BASE-F is the fiber-optic version of IEEE 802.3.
- 10BASE-F includes 10BASE-FB (backbone), 10BASE-FL (concentrator-to-station links), and 10BASE-FP (passive star).
- FDDI is a token-passing network originally designed to use fiber optics.
- FDDI operates at 100 Mbps, allows 100 attached nodes, and has a network length of 100 km.
- FDDI uses two counter-rotating rings for fault tolerance.

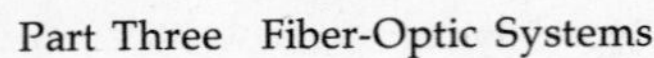

- ESCON was the first fiber-optic channel for mainframe computers offered as standard equipment.
- Fiber Channel is an industry-standard channel specification with speeds up to 1.062 Gbaud.
- Sonet is a international standard for optical digital telecommunications offering data rates to 10 Gbps.
- Fiber optics in the local loop will bring fiber either to your curb or to your home.
- An erbium-doped fiber amplifier is a type of self-amplifying fiber used to amplify 1550-nm signal by pumping with a shorter wavelength.
- Solitons are optical pulses that do not disperse as they propagate and, therefore, promise *very* long distance repeaterless transmission.

REVIEW QUESTIONS

1. Name the three most common LANs in use today and give their maximum transmission rate.
2. Name four types of cable used with IEEE 802.3 networks.
3. What is the most common fiber size used in LANs?
4. Define FDDI in terms of name, transmission rate, transmission distance, and number of nodes.
5. Define SAS and DAS as they relate to FDDI.
6. What is the purpose of the Director in the IBM ESCON system?
7. In Sonet, what would be the transmission rate of an OC-7 carrier?
8. What is the minimum number of bytes in an FDDI data frame?
9. What is the main reason for using a FOIRL transceiver in an IEEE 802.3 network?
10. What are the three variations of a 10BASE-F network?

CHAPTER 16

Introduction to Test and Other Equipment

This chapter provides a brief look at some of the equipment commonly used in the field to install, inspect, and maintain fiber-optic systems. It also discusses methods to measure fiber attenuation and connector insertion loss. The equipment discussed is the optical power meter, optical time-domain reflectometer, fusion splicer, polishing machine, inspection microscope, and a few hand tools. Our discussion will be simplified and clarified by looking at specific examples of each.

FIBER-OPTIC TESTING

A great many tests must be performed on optical fibers. A fiber manufacturer must test a fiber to determine the characteristics by which the fiber will be specified. As a quality control measure during manufacture of fibers, the manufacturer must constantly test the fibers to ensure that they meet the specifications. Among such tests are the following:

- Core diameter
- Cladding diameter
- NA
- Attenuation
- Refractive index profile
- Tensile strength

Other tests performed on fibers or on fiber-optic cables concern their mechanical and environment characteristics. Mechanical tests such as impact resistance, tensile loading, and crush resistance test the cable's ability to withstand physical and mechanical stresses. Environmental tests evaluate the changes in attenuation under extremes of temperature, repeated changes (cycling) of temperature, and humidity.

Fiber and cables are specified as the result of such tests. Engineers pick cables suited to their applications on the basis of specifications. Manufacturers of other fiber-optic components—sources, detectors, connectors, and couplers—also test to characterize and specify their offerings.

Other tests are performed during and after installation of a fiber-optic link. These tests ensure that the installed system will meet performance requirements. Splices, for example, must not exceed certain loss values. During installation, each splice must be tested. If the loss is unacceptably high, the fibers must be respliced. This chapter describes how such a test is done with optical time reflectometry.

OPTICAL POWER METER

The *optical power meter* is analogous to the volt-ohm-milliammeter (VOM) used in electronics. Figure 16–1 shows a power meter. The meter reads optical power levels, either in absolute levels of dBm and dBμ or in relative units of dB.

The meter itself is completely electronic. Sensors that plug into the unit contain the detector and perform the optical-to-electrical conversion. Different sensors are available for use at different power levels and operating wavelengths from 400 to 1800 nm. Adapters permit bare fibers or a variety of popular connectors to be connected to a sensor.

The range of the meter, with different sensor heads, is –80 dBm (10 pW) to +33 dBm (2 W). Resolution is selectable at 0.1 or 0.01 dB. Resolution determines whether the reading will be one or two decimal places. An output jack, providing an adjustable, regulated current, allows plug-in LEDs to provide the light source for measurements.

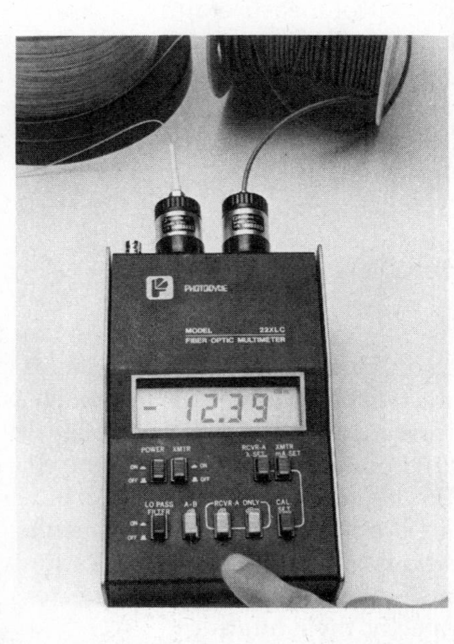

FIGURE 16–1 Optical power meter (Photo courtesy of Photodyne)

The meter can be used for a number of measurements. We will describe its use for fiber attenuation measurements and connector insertion loss measurements. But first, we must discuss mode control, an important factor to test results and repeated tests.

MODE CONTROL

Chapter 11 discussed the influence of launch and receive conditions on connector insertion loss. Such launch and receive conditions result from the mode distribution in fibers. Long launch or receive conditions mean that the equilibrium mode distribution exists in the fiber. For repeatable measurements, launch conditions especially must be controlled. Most often, long launch, or EMD, conditions are preferred. The Electronics Industry Association (EIA) recommends a 70/70 launch: 70% of the fiber core diameter and 70% of the fiber NA should be filled. This recommendation corresponds to EMD in a graded-index fiber. EMD can be reached by three approaches: optical, filtering, and long fiber length.

The optical approach uses collimating and focusing lenses and optical apertures to direct the light into the center of the fiber core. Although the result is precise, this approach is only practical in a laboratory equipped with an optics bench.

The long-fiber approach uses a fiber long enough to reach EMD regardless of the light injected into it. Even though the source overfills or underfills the fiber, EMD is reached over distance. The disadvantage is that the required fiber length may be several kilometers.

Filtering presents the most practical approach. A mode filter causes mode mixing to simulate EMD in a short length of fiber. A half-in. diameter mandrel around which a fiber is wrapped five or six times is a standard approach. The five wraps cause mode mixing so that high-order modes in an overfilled fiber are lost.

The mandrel wrap, however, does not remove cladding modes. Cladding modes will propagate a short distance in multimode fiber, and it may be necessary to remove them if the fiber being used is short. A cladding mode stripper is a material with a refractive index greater than the cladding. A bath of glycerin, oil, or other suitable liquid will serve. The buffer coating must be removed to expose the cladding, which is immersed in the mode stripper. Because of the greater refractive index of the stripper, light is no longer bound in the cladding. It passes into the stripper.

FIBER LOSS MEASUREMENTS

Losses in a fiber-optic cable can be measured in two ways. The cutback method, shown in Figure 16–2, uses a single fiber. First, power through the fiber is measured. A section of fiber is then cut off, and the power through the

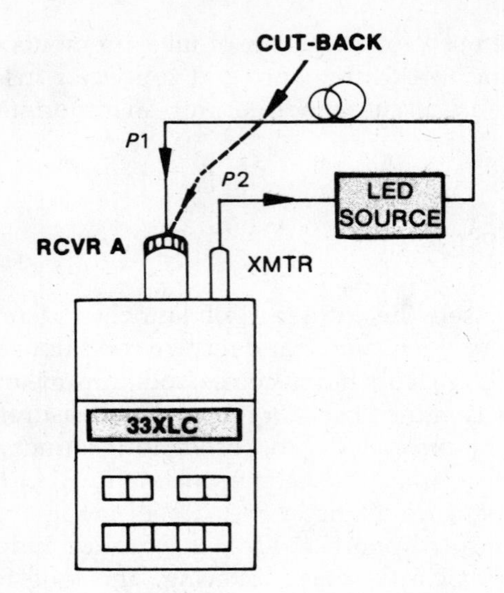

FIGURE 16–2 Attenuation measurement by cutback technique (Courtesy of Photodyne)

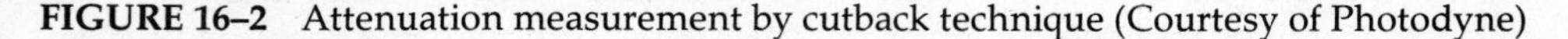

remaining fiber is measured. The fiber loss in decibels per kilometer can then be determined by

$$\text{loss} = \frac{P_2 - P_1}{L}$$

where P_1 is the dBm reading from the first measurement, P_2 is the dBm reading from the second measurement, and L is the difference, in kilometers, in the two cable lengths. Notice that subtracting one dBm unit from another yields dB. Changing the units of the difference in fiber lengths allows loss per unit length to be measured in other units.

A second method of measuring attenuation in an optical fiber is to compare the power through the cable under test to the power of a known reference cable. The power through the cable under test is measured in absolute units on the meter. Next, power through the reference cable is measured.

A power meter can also be used to check the output power from a transmitter during quality control or troubleshooting.

INSERTION LOSS TESTS

Chapter 11 discussed insertion loss testing for connectors and splices briefly. In essence, the test measures the power through a length of fiber. The fiber is then cut near the center, and the connector or splice is applied to the exposed ends. Power through the cable is again measured. The difference between the two readings is the insertion loss contributed by the connector or splice. Here we wish to emphasize the importance of mode control in the test. Since losses will vary, depending on the launch and receive conditions, controlling these conditions with mode filters and cladding mode strippers is essential to understandable, meaningful, and repeatable results.

A main drawback to the use of an optical power meter in many applications is that both ends of the fiber must be available. An optical time-domain reflectometer allows testing when only one end of the fiber is conveniently available.

TIME AND FREQUENCY DOMAINS

A signal can be described in terms of the time domain or the frequency domain. This book has used both domains without bothering to distinguish between them. We described the limitations on a fiber's information-carrying capacity in terms of both bandwidth and dispersion. Bandwidth is in the frequency domain; rise-time dispersion is in the time domain. Analog engineers, dealing with analog signals in the frequency domain, talk in terms of frequencies. Digital engineers, dealing with pulses, use the time domain and talk in terms of rise times and pulse widths. We have given a simple method of relating the frequency and time domain in the equations relating bandwidth and rise time.

$$t_r = \frac{0.35}{\text{BW}}$$

$$\text{BW} = \frac{0.35}{t_r}$$

OPTICAL TIME-DOMAIN REFLECTOMETER

As its name implies, the *optical time-domain reflectometer* allows evaluation of an optical fiber in the time domain. Figure 16–3 shows an optical time-domain reflectometer.

Optical time-domain reflectometry (OTDR) relies on the backscattering of light that occurs in an optical fiber. Some of the light entering the fiber will be reflected back to the light. *Backscattered* light is that light that reaches the input end of a fiber. Backscattering results from Rayleigh scattering and Fresnel reflections. Rayleigh scattering, remember, is scattering caused by the refractive displacement due to density and compositional variations in the fiber. In a quality fiber, the scattered light can be assumed to be evenly distributed with length. Fresnel reflections occur because of changes in refractive index at connections, splices, and fiber ends. A portion of the Rayleigh scattered light and Fresnel reflected light reaches the input end as backscattered light.

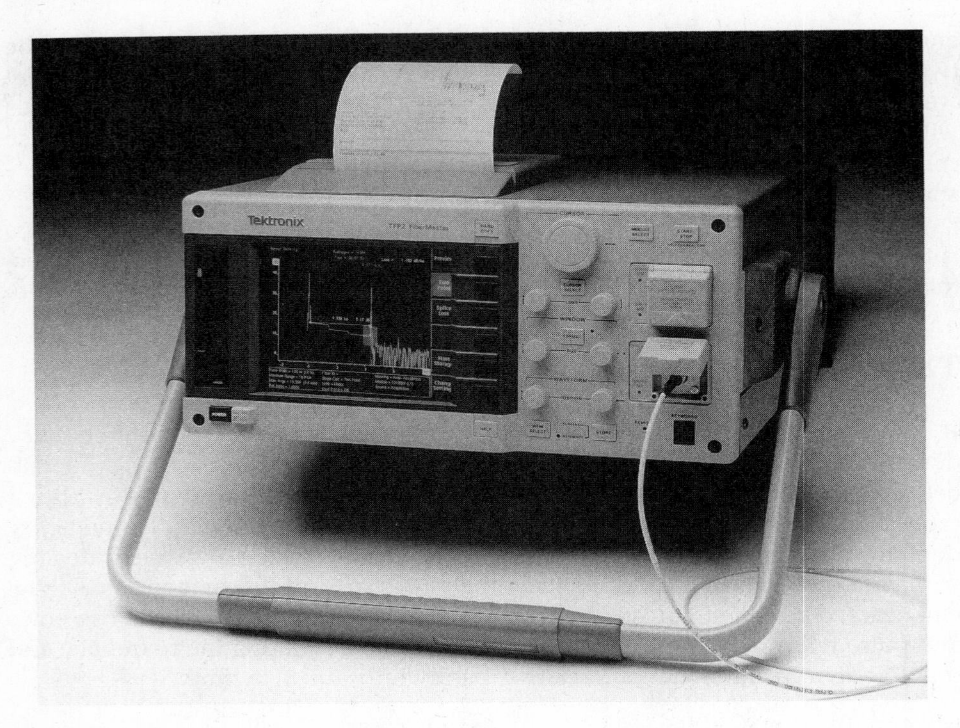

FIGURE 16–3 Optical time-domain reflectometer (Photo courtesy of Tektronix, Inc.)

Figure 16–4 shows a simple block diagram of an OTDR unit. Its main parts are a light source, a beamsplitter, a photodetector, and an oscilloscope. A short, high-powered pulse is injected through the beamsplitter into the fiber. This light is then backscattered as it travels through the fiber. The beamsplitter directs the backscattered light to the photodetector. The amplified output of the detector serves as the vertical input to the oscilloscope. Because the power to the detector is extremely small, repeated measurements are made by the electronics of the ODTR; the SNR is improved by averaging the readings, after which the results are displayed.

The OTDR screen displays time horizontally and power vertically. Fiber attenuation appears as a line decreasing from left (the input end of the fiber) to right (the output end). Both the input and the backscattered light attenuate over distance, so the detected signal becomes smaller over time. A connector, splice, fiber end, or abnormality in the fiber appears as an increase in power on the screen, since backscattering from Fresnel reflections will be greater than backscattering from Rayleigh scattering. The quality of a splice can be evaluated by the amount of backscattering: Greater backscattering means a higher-loss splice. A connector

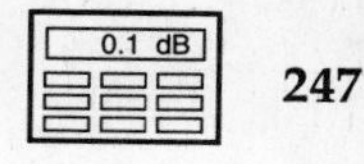

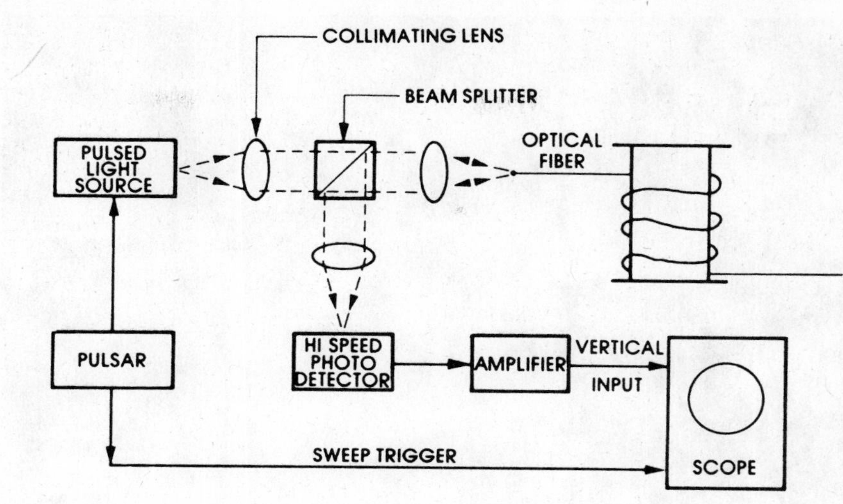

FIGURE 16–4 OTDR block diagram (Courtesy of Photodyne)

shows both a power increase from reflection and a power drop from loss. The degree of loss indicates the quality of the connection. Figure 16–5 shows a typical display. Notice that it is similar to the power budget diagram in Chapter 13.

Light travels through a fiber at a speed of about 5 ns/m, depending on the refractive index of the core. Time can be correlated to distance by

$$D = \frac{ct}{2n}$$

where D is distance along the fiber, c is the speed of light, t is the round-trip travel time of the input pulse, and n is the core's average refractive index. Most OTDRs use a cursor to mark a horizontal point on the trace and display the distance in terms of both time and physical distance. One can, for example, determine the distance to a splice with a great degree of accuracy, typically to within a foot. The OTDR measures the distance along the fiber, not necessarily the length of the cable, however. If the fiber is stranded around a central core, the actual fiber length will be somewhat longer than the cable length.

The lengths at which the OTDR can be used depend on two things. First is the dynamic range of the units, which sets the minimum and maximum power to the detector. Beyond that, the length is determined by fiber attenuation and losses at splices and connectors. OTDR dynamic range and power losses within the fiber system set the length possible before the backscattered light becomes too weak to be detected. With typical low-loss fibers used in long-distance telecommunications, the OTDR can be used on lengths of 20 to 40 km.

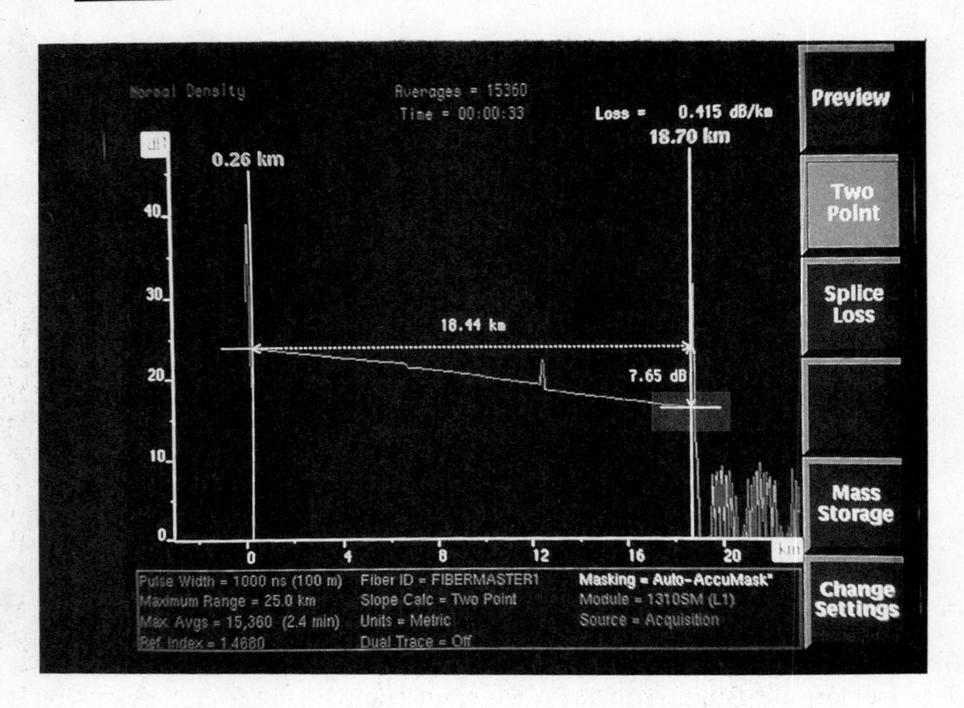

FIGURE 16–5 OTDR screen (Photo courtesy of Tektronix, Inc.)

A sophisticated OTDR unit offers the technician a great deal of information. The screen in Figure 16–5 shows a loss of 7.65 dB over an 18.44-km link span, for an average of 0.415 dB/km. The large bump at about 13 km represents a reflection, as from a splice or connector; the small bump at 7 km is another splice, but with minimal loss and very little reflection. Notice also that the bottom of the screen shows conditions of the setup. Two items are of interest. First, source module is 1310SM, meaning the OTDR source is a laser operating at 1310 nm into a single-mode fiber. Second, the refractive index of the fiber is given as 1.4680. The OTDR uses this value to calculate distances based on time. If the refractive index value is wrong, the distances will be somewhat incorrect.

The OTDR in Figure 16–3 uses plug-in modules to permit operation at 850, 1310, and 1550 nm on both single-mode and multimode fibers. It can operate on links over 100 km long. Because the unit uses an internal processor to analyze and display measurements, it can store waveforms. It also allows you to change the resolution from a long view of the overall link span to a closeup of a particular event. You could, for example, zoom in for a close inspection of the reflection in Figure 16–4 to measure the reflection loss.

The OTDR can tell you a great deal about a link and allow you to inspect various sections of a link in detail.

OTDR principles have been adapted to less expensive equipment. An optical fault finder, for example, uses time-domain reflectometry to measure the distance to a fault. Typically, however, it provides only a simple numerical readout of the distance to the fault. The advantage is a low-cost, compact unit, often handheld and battery operated.

USES OF OTDR

OTDR has many uses in research and development and in manufacturing. Here we will discuss three field uses: loss per unit length, splice and connector evaluation, and fault location. These three uses are very important during installation and for system maintenance. Loss per unit length and evaluation of connectors and splices can also be done with an optical power meter. The advantage of OTDR in field applications involving any appreciable length is that only one end of the fiber must be available. The OTDR unit can be placed in a sheltered environment and kept out of manholes. In applications involving short lengths of fibers, such as offices, the power meter might be more convenient.

Loss Per Unit Length

The loss budget of a fiber-optic installation assumes certain fiber attenuation, which is loss per unit length. OTDR can measure attenuation before and after installation. Measurements before installation ensure that all fibers meet specified limits. Measurements after installation check for any increases in attenuation that could result from such things as bends or unexpected loads.

Splice and Connector Evaluation

Here OTDR can be used during the installation processes to ensure that all splices and connector losses are within acceptable limits. After the splice or connector is applied, an OTDR reading checks the loss. If it is unacceptable, another connector or splice is applied.

Some splices are tunable. By rotating one half of the splice in relation to the other, losses from lack of symmetry in the fiber or splice can be minimized. As the splice is rotated, the OTDR display is monitored to determine the point of lowest loss and maximum transmission.

The OTDR unit and operator are usually in a location such as a central telephone office. Communication with the splicer is by radio or telephone.

Fault Location

Faults, such as broken fibers or splices, may occur during or after installation. A telephone line will be broken two or three times during its 30-year lifetime.

The location of the fault may not be apparent if the cable is buried or in a conduit or if only the fiber and not the cable is broken. OTDR provides a useful method of locating the fault accurately and thereby saving much time and expense. It certainly beats digging up several kilometers of cable unnecessarily.

FUSION SPLICER

Chapter 11 described a fusion splice. A fusion splicer (Figure 16–6) uses an electric arc to heat fibers to about 2000 °C. The fibers melt, are pushed together, and fuse together as they cool.

While fusion splices are used with multimode fibers, single-mode fibers present an additional difficulty because of their core diameter. With multimode fibers, it is sufficient to align the claddings to achieve low splice loss. Single-mode fibers, however, demand that the cores be aligned.

Achieving the best fusion is a complex matter involving not only precise alignment of fibers, but careful application of the correct arc discharge power and timing. An incorrect discharge, for example, can deform the cores. Splicers have adjustable power and time to accommodate different fibers and application needs.

The earliest splicers required a great degrees of operator skill to align the fibers, often visually through a microscope. Often power through the fibers was measured as the fibers were being aligned. At the point of maximum power at the junction, the fibers were locked in place and fused. Such power monitoring

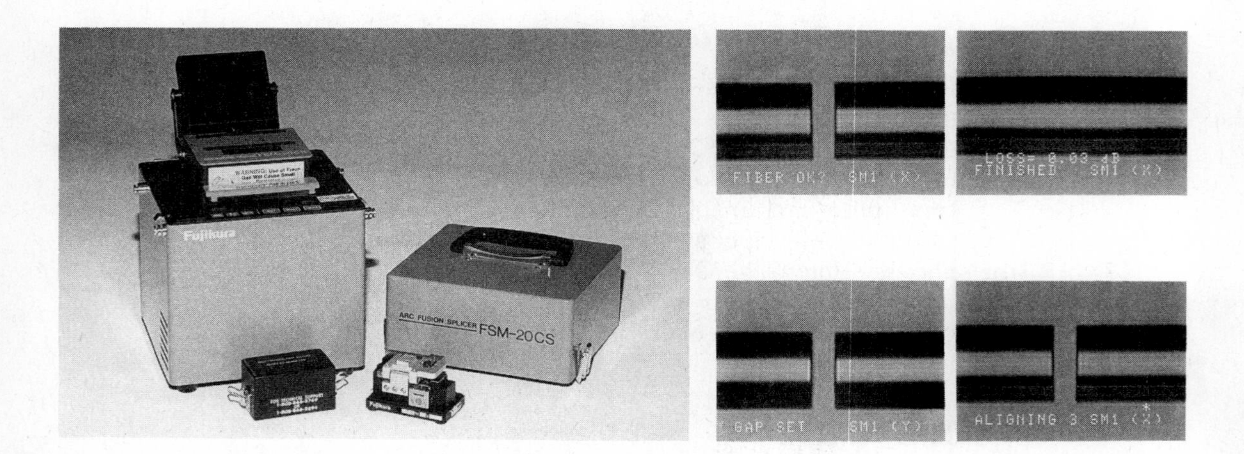

FIGURE 16–6 Fusion splicer (Photo courtesy of Alcoa Fujikura Ltd.)

could be done remotely with an OTDR or locally by techniques that injected a small amount of light into the cladding at a bend in the fiber.

Micropositioners hold the fibers and allow them to be precisely aligned three dimensionally along the x, y, and z axes, either manually or automatically.

Today's splicers have a high degree of computer control and analysis to automate the fusion process. One example—known as profile alignment system—is shown in Figure 16–7. Collimated light is reflected off a mirror, through the fiber at right angles to the axis, and into a video camera. The video camera connects both to a screen for the operator and to a computer that analyzes the images.

Because of the way light is refracted by the different refractive indices of the fiber, the cladding will appear dark and the core will appear light. The splicer's computer analyzes the images to locate the centerlines of the cores. The computer then moves the fibers into alignment. Notice that the camera can move to analyze the fiber on two perpendicular planes. Once the fibers are aligned, the computer will estimate the loss of the splice. If the value is not acceptable, the operator can clean the fibers or recleave them. Once the loss is within acceptable limits, the operator initiates the fuse cycle. After the splice is complete, the operator can view the alignment of the cores.

The biggest danger in fusion splicing is the fibers shifting along their axis during fusion due to the surface tension of the molten glass. The splicer, by

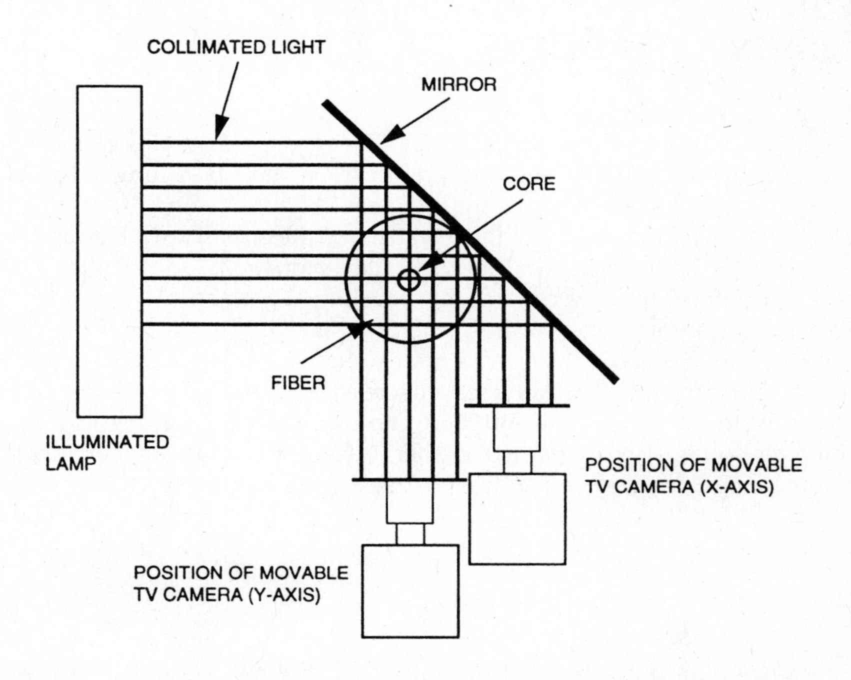

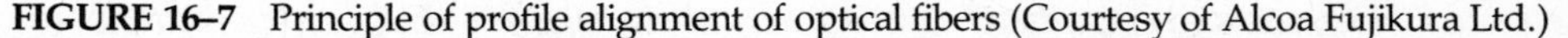

FIGURE 16–7 Principle of profile alignment of optical fibers (Courtesy of Alcoa Fujikura Ltd.)

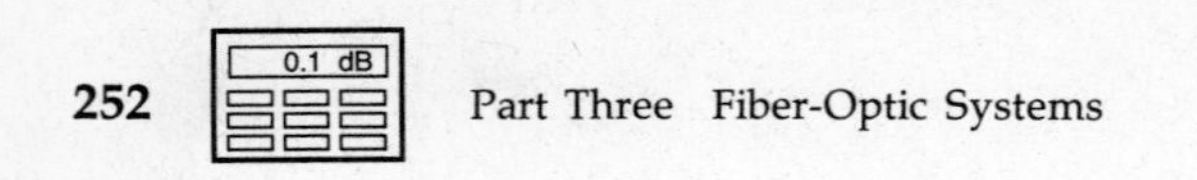

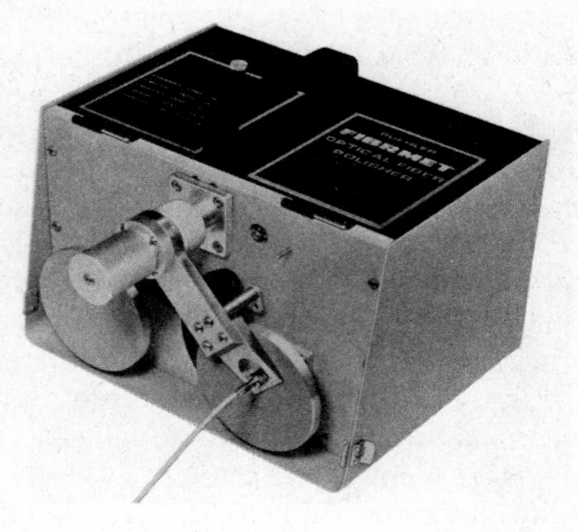

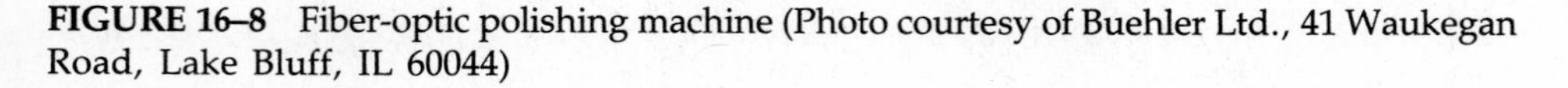

FIGURE 16–8 Fiber-optic polishing machine (Photo courtesy of Buehler Ltd., 41 Waukegan Road, Lake Bluff, IL 60044)

knowing the parameters of the fiber and the arc discharge, can compensate automatically for axis shift.

POLISHING MACHINE

The polishing machine shown in Figure 16–8 partially automates the polishing process to produce consistent end finishes. The machine contains two polishing platens: one for rough polishing and one for final, fine polishing. Replaceable circular polishing disks adhere to the platens.

Connectors held in precision adapters attach to the sweep arm. The arm moves up and down at 12 sweeps per minute while the platens spin. The adapters contain jeweled stops that precisely control the polishing depth.

Because connectors are polished, the fibers are not often carefully cleaved flush against the connector face. As a result, the fiber often protrudes from the connector. The drive mechanism uses a decreasing pneumatic pressure to make soft starts during polishes. The soft start prevents damage to the polishing disk and, more importantly, fractures to the fiber. Polishing pressure is then increased. Approximate polishing times are 30 s for rough and 45 s for fine. After polishing, the finish should be inspected under a microscope or magnifier.

INSPECTION MICROSCOPE

Portable inspection microscopes (Figure 16–9) allow closeup inspection of cleaved fiber ends and polishing finishes in the field. Closeups inspection is an

FIGURE 16–9 Inspection microscope (Illustration courtesy of AMP Incorporated)

important quality control procedure for evaluating splices and connectors. It is much simpler than OTDR, although it can be used in conjunction with OTDR.

The microscope allows both customary through-lens illumination and oblique illumination that reveals pits, scratches, and polishing patterns. Connectors or fibers can be rotated to allow finishes to be viewed across the end, straight on, or at angles in between. Different adapters are required for different connectors and bare fibers. Interchangeable eyepiece lenses and objective lenses allow a magnification range from 40X to 800X.

INSTALLATION KITS

Installation of fiber-optic cables requires many tools, some of which are common and some of which are special to fiber optics and the products being used. An installation kit, such as shown in Figure 16–10, contains all the tools commonly needed to split open multifiber cables, prepare fibers, apply connectors, polish connectors, and inspect the finish.

A partial list of the kit contents includes

- Screwdriver and nutdriver for preparing organizers
- Side-cutting pliers and utility knife for removing cable sheaths
- Tape measure for measuring cable and fiber stripping lengths
- Cable stripper for removing jacketing
- Fiber strippers for removing buffer coatings
- Scissors for trimming Kevlar strength members
- Roller for mixing epoxy
- Scribe tool for trimming fibers
- Crimp tool for crimping connectors to fibers

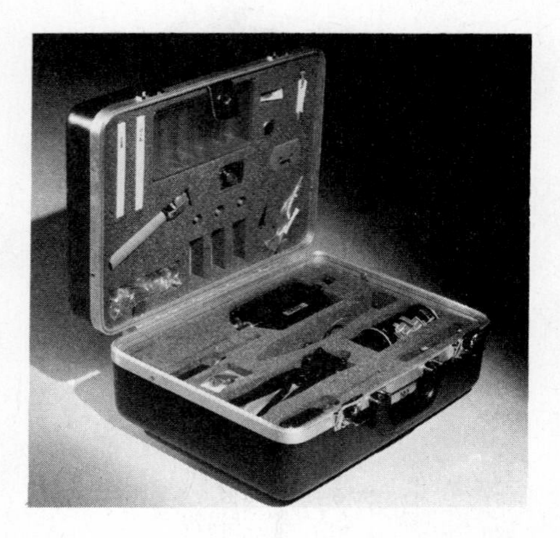

FIGURE 16–10 Installation kit (Photo courtesy of AMP Incorporated)

- Polishing materials
- Duster (a canned air for cleaning fibers)
- Heat gun for applying heat-shrink tubing
- Index-matching fluid for lowering or eliminating loss from Fresnel reflections in a fiber-to-fiber interconnection
- Inspection microscope for inspecting fiber end finishes

Such kits provide the proper tools for the job in one convenient location.

SUMMARY

- An optical power meter allows optical power levels to be measured.
- Mode control is often important to repeatable test results.
- An optical time reflectometer is used to evaluate connectors and splices, locate faults, and measure loss per unit length in a fiber.
- A fusion splicer fuses fiber in fiber-to-fiber interconnections.
- A microscope allows end finishes to be inspected.
- A polishing machine produces fine, consistent end finishes.

REVIEW QUESTIONS

1. Name the electrical test equipment to which an optical power meter is most easily compared.
2. Describe a method of measuring fiber loss with a power meter.

3. List three methods of obtaining EMD in a fiber under test.
4. For each of the following, say whether the time domain or the frequency domain applies:
   A. Rise time
   B. Bandwidth
   C. Pulse width
   D. 500 MHz
   E. 20 ns
   F. Dispersion
   G. Source spectral width
   H. ODTR
5. Name two factors that limit the distance at which an OTDR unit can be used.
6. List three uses of OTDR.
7. Sketch an OTDR display for a 20-km link with a splice at 5 km and connectors at 10 and 15 km.
8. Besides performing the splice itself, name two other essential activities required of a good splice and the fusion splicer described in this chapter allows to be performed.
9. How many polishing stages are there in polishing a typical connector? What is the function of each stage?

GLOSSARY

Absorption Loss of power in an optical fiber, resulting from conversion of optical power into heat and caused principally by impurities, such as transition metals and hydroxyl ions, and also by exposure to nuclear radiation.

Acceptance angle The half-angle of the cone within which incident light is totally internally reflected by the fiber core. It is equal to arcsin (NA).

AM Amplitude modulation.

Amplitude modulation A transmission technique in which the amplitude of the carrier is varied in accordance with the signal.

Angular misalignment The loss of optical power caused by deviation from optimum alignment of fiber to fiber or fiber to waveguide.

APD Avalanche photodiode.

Attenuation A general term indicating a decrease in power from one point to another. In optical fibers, it is measured in decibels per kilometer at a specified wavelength.

Attenuation-limited operation The condition in a fiber-optic link when operation is limited by the power of the received signal (rather than by bandwidth or by distortion).

Avalanche photodiode A photodiode that exhibits internal amplification of photocurrent through avalanche multiplication of carriers in the junction region.

Backscattering The return of a portion of scattered light to the input end of a fiber; the scattering of light in the direction opposite to its original propagation.

Bandwidth A range of frequencies.

Bandwidth-limited operation The condition in a fiber-optic link when bandwidth, rather than received optical power, limits performance. This condition is reached when the signal becomes distorted, principally by dispersion, beyond specified limits.

Baseband A method of communication in which a signal is transmitted at its original frequency without being impressed on a carrier.

Baud A unit of signaling speed equal to the number of signal symbols per second, which may or may not be equal to the data rate in bits per second.

Beamsplitter An optical device, such as a partially reflecting mirror, that splits a beam of light into two or more beams and that can be used in fiber optics for directional couplers.

Bend loss A form of increased attenuation in a fiber that results from bending a fiber around a restrictive curvature (a macrobend) or from minute distortions in the fiber (microbends).

BER Bit-error rate.

Bit A binary digit, the smallest element of information in binary system. A 1 or 0 of binary data.

Bit-error rate The ratio of incorrectly transmitted bits to correctly transmitted bits.

Broadband A method of communication in which the signal is transmitted by being impressed on a higher-frequency carrier.

Buffer A protective layer over the fiber, such as a coating, an inner jacket, or a hard tube.

Buffer coating A protective layer, such as an acrylic polymer, applied over the fiber cladding for protective purposes.

Buffer tube A hard plastic tube, having an inside diameter several times that of a fiber, that holds one or more fibers.

Bus network A network topology in which all terminals are attached to a transmission medium serving as a bus.

Byte A unit of 8 bits.

Carrier sense multiple access with collision detection A technique used to control the transmission channel of a local area network to ensure that there is no conflict between terminals that wish to transmit.

Centro-symmetrical reflective optics An optical technique in which a concave mirror is used to control coupling of light from one fiber to another.

Channel A communications path or the signal sent over the channel. Through multiplexing several channels, voice channels can be transmitted over an optical channel.

Cladding The outer concentric layer that surrounds the fiber core and has a lower index of refraction.

Cladding mode A mode confined to the cladding; a light ray that propagates in the cladding.

Concentrator A multiport repeater.

Connector A device for making connectable/disconnectable connections of a fiber to another fiber, source, detector, or other devices.

Core The central, light-carrying part of an optical fiber; it has an index of refraction higher than that of the surrounding cladding.

Coupler A multiport device used to distribute optical power.

CSMA/CD Carrier sense multiple access with collision detection.

CSR Centro-symmetrical reflective optics.

Cutoff wavelength For a single-mode fiber, the wavelength above which the fiber exhibits single-mode operation.

Dark current The thermally induced current that exists in a photodiode in the absence of incident optical power.

Data rate The number of bits of information in a transmission system, expressed in bits per second (bps), and which may or may not be equal to the signal or baud rate.

dB Decibel.

dBm Decibel referenced to a milliwatt.

dBμ Decibel referenced to a microwatt.

Decibel A standard logarithmic unit for the ratio of two powers, voltages, or currents. In fiber optics, the ratio is power.

$$\text{dB} = 10 \log_{10} \left(\frac{P_1}{P_2} \right)$$

Detector An optoelectronic transducer used in fiber optics for converting optical power to electric current. In fiber optics, usually a photodiode.

Diameter-mismatch loss The loss of power at a joint that occurs when the transmitting half has a diameter greater than the diameter of the receiving half. The loss occurs when coupling light from a source to fiber, from fiber to fiber, or from fiber to detector.

Dichroic filter An optical filter that transmits light selectively according to wavelength.

Diffraction grating An array of fine, parallel, equally spaced reflecting or transmitting lines that mutually enhance the effects of diffraction to concentrate the diffracted light in a few directions determined by the spacing of the lines and by the wavelength of the light.

Dispersion A general term for those phenomena that cause a broadening or spreading of light as it propagates through an optical fiber. The three types are modal, material, and waveguide.

Distortion-limited operation Generally synonymous with bandwidth-limited operation.

Duplex cable A two-fiber cable suitable for duplex transmission.

Duplex transmission Transmission in both directions, either one direction at a time (half duplex) or both directions simultaneously (full duplex).

Duty cycle In a digital transmission, the ratio of high levels to low levels.

EDFA Erbium-doped fiber amplifier.

8B/10B encoding A signal modulation scheme in which either four bits are encoded into a five-bit word or eight bits are encoded in a 10-bit word to ensure that too many consecutive zeroes do not occur; used in ESCON and Fiber Channel.

802.3 network A 10-Mbps CSMA/CD bus-based network; commonly called Ethernet.

802.5 network A token-passing ring network operating at 4 or 16 Mbps.

Electromagnetic interference Any electrical or electromagnetic energy that causes undesirable response, degradation, or failure in electronic equipment. Optical fibers neither emit nor receive EMI.

EMD Equilibrium mode distribution.

EMI Electromagnetic interference.

Equilibrium mode distribution The steady modal state of a multimode fiber in which the relative power distribution among modes is independent of fiber length.

Erbium-doped fiber amplifier A type of fiber that amplifies 1550-nm optical signals when pumped with a 980- or 1480-nm light source.

ESCON An IBM channel control system based on fiber optics.

Excess loss In a fiber-optic coupler, the optical loss from that portion of light that does not emerge from the nominally operational ports of the device.

Extrinsic loss In a fiber interconnection, that portion of loss that is not intrinsic to the fiber but is related to imperfect joining, which may be caused by the connector or splice.

Fall time The time required for the trailing edge of a pulse to fall from 90% to 10% of its amplitude; the time required for a component to produce such a result. "Turn off time." Sometimes measured between the 80% and 20% points.

FDDI Fiber Distributed Data Interface.

Fiber Channel An industry-standard specification for computer channel communications over fiber optics and offering transmission speeds from 132 Mbaud to 1062 Mbaud and transmission distances from 1 to 10 km.

Fiber Distributed Data Interface network A token-passing ring network designed specifically for fiber optics and featuring dual counterrotating rings and 100 Mbps operation.

Fiber-Optic Interrepeater Link A standard defining a fiber-optic link between two repeaters in an IEEE 802.3 network.

FM Frequency modulation.

FOIRL Fiber-optic interrepeater link.

4B/5B Encoding A signal modulation scheme in which groups of four bits are encoded and transmitted in five bits in order to guarantee that no more than three consecutive zeroes ever occur; used in FDDI.

Frequency modulation A method of transmission in which the carrier frequency varies in accordance with the signal.

Fresnel reflection The reflection that occurs at the planar junction of two materials having different refractive indices; Fresnel reflection is not a function of the angle of incidence.

Fresnel reflection loss Loss of optical power due to Fresnel reflections.

Fused coupler A method of making a multimode or single-mode coupler by wrapping fibers together, heating them, and pulling them to form a central unified mass so that light on any input fiber is coupled to all output fibers.

Gap loss Loss resulting from the end separation of two axially aligned fibers.

Graded-index fiber An optical fiber whose core has a nonuniform index of refraction. The core is composed of concentric rings of glass whose refractive indices decrease from the center axis. The purpose is to reduce modal dispersion and thereby increase fiber bandwidth.

Ground-loop noise Noise that results when equipment is grounded at ground points having different potentials and thereby creating an unintended current path. The dielectric of optical fibers provides electrical isolation that eliminates ground loops.

IDP Integrated detector/preamplifier.

Index-matching material A material, used at optical interconnection, having a refractive index close to that of the fiber core and used to reduce Fresnel reflections.

Index of refraction The ratio of the velocity of light in free space to the velocity of light in a given material. Symbolized by n.

Insertion loss The loss of power that results from inserting a component, such as a connector or splice, into a previously continuous path.

Integrated detector/preamplifier A detector package containing a pin photodiode and transimpedance amplifier.

ISO International Standard Organization.

LAN Local area network.

Laser A light source producing, through stimulated emission, coherent, near monochromatic light. Lasers in fiber optics are usually solid-state semiconductor types.

Lateral displacement loss The loss of power that results from lateral displacement from optimum alignment between two fibers or between a fiber and an active device.

LED Light-emitting diode.

Light-emitting diode A semiconductor diode that spontaneously emits light from the pn junction when forward current is applied.

Local area network A geographically limited network interconnecting electronic equipment.

Material dispersion Dispersion resulting from the different velocities of each wavelength in an optical fiber.

MFD Mode field diameter.

Misalignment loss The loss of power resulting from angular misalignment, lateral displacement, and end separation.

Modal dispersion Dispersion resulting from the different transit lengths of different propagating modes in a multimode optical fiber.

Mode In guided-wave propagation, such as through a waveguide or optical fiber, a distribution of electromagnetic energy that satisfies Maxwell's equations and boundary conditions. Loosely, a possible path followed by light rays.

Mode coupling The transfer of energy between modes. In a fiber, mode coupling occurs until EMD is reached.
Mode field diameter The diameter of optical energy in a single-mode fiber. Because the MFD is greater than the core diameter, MFD replaces core diameter as a practical parameter.
Mode filter A device used to remove high-order modes from a fiber and thereby simulate EMD.
Modulation The process by which the characteristic of one wave (the carrier) is modified by another wave (the signal). Examples include amplitude modulation (AM), frequency modulation (FM), and pulse-coded modulation (PCM).
Multimode fiber A type of optical fiber that supports more than one propagating mode.
Multiplexing The process by which two or more signals are transmitted over a single communications channel. Examples include time-division multiplexing and wavelength-division multiplexing.

NA Numerical aperture.
NA-mismatch loss The loss of power at a joint that occurs when the transmitting half has an NA greater than the NA of the receiving half. The loss occurs when coupling light from a source to fiber, from fiber to fiber, or from fiber to detector.
Numerical aperture The "light-gathering ability" of a fiber, defining the maximum angle to the fiber axis at which light will be accepted and propagated through the fiber. NA = $\sin \theta$, where θ is the acceptance angle. NA is also used to describe the angular spread of light from a central axis, as in exiting a fiber, emitting from a source, or entering a detector.

Open Standard Interconnect A seven-layer model defined by ISO for defining a communication network.
Optical time-domain reflectometry A method of evaluating optical fibers based on detecting backscattered (reflected) light. Used to measure fiber attenuation, evaluate splice and connector joints, and locate faults.
OSI Open Standards Interconnect.
OTDR Optical time-domain reflectometry.

PC Physical contact.
PCM Pulse-coded modulation.
PCS Plastic-clad silica.
Photodetector An optoelectronic transducer, such as a pin photodiode or avalanche photodiode.
Photodiode A semiconductor diode that produces current in response to incident optical power and used as a detector in fiber optics.
Photon A quantum of electromagnetic energy. A "particle" of light.
Physical contact connector A connector designed with a radiused tip to ensure physical contact of the fibers and thereby increase return reflection loss.

Pigtail A short length of fiber permanently attached to a component, such as a source, detector, or coupler.

Pin photodiode A photodiode having a large intrinsic layer sandwiched between p-type and n-type layers.

Pistoning The movement of a fiber axially in and out of a ferrule end, often caused by changes in temperature.

Plastic-clad silica fiber An optical fiber having a glass core and plastic cladding.

Plastic fiber An optical fiber having a plastic core and plastic cladding.

Plenum The air space between walls, under structural floors, and above drop ceilings, which can be used to route intrabuilding cabling.

Plenum cable A cable whose flammability and smoke characteristics allow it to be routed in a plenum area without being enclosed in a conduit.

Pulse-coded modulation A technique in which an analog signal, such as a voice, is converted into a digital signal by sampling the signal's amplitude and expressing the different amplitudes as a binary number. The sampling rate must be twice the highest frequency in the signal.

Pulse spreading The dispersion of an optical signal with time as it propagates through an optical fiber.

Quantum efficiency In a photodiode, the ratio of primary carriers (electron-hole pairs) created to incident photons. A quantum efficiency of 70% means 7 out of 10 incident photons create a carrier.

Rayleigh scattering The scattering of light that results from small inhomogeneities in material density or composition.

Regenerative repeater A repeater designed for digital transmission that both amplifies and reshapes the signal.

Repeater A device that receives, amplifies (and perhaps reshapes), and retransmits a signal. It is used to boost signal levels when the distance between repeaters is so great that the received signal would otherwise be too attenuated to be properly received.

Responsivity The ratio of a photodetector's electrical output to its optical input in amperes/watt.

Return reflection Reflected optical energy that propagates backward to the source in an optical fiber.

Return reflection loss The attenuation of reflected light; high return loss is desirable, especially in single-mode fibers.

Ring network A network topology in which terminals are connected in a point-to-point serial fashion in an unbroken circular configuration.

Rise time The time required for the leading edge of a pulse to rise from 10% to 90% of its amplitude; the time required for a component to produce such a result. "Turn-on time." Sometimes measured between the 20% and 80% points.

Sensitivity For a fiber-optic receiver, the minimum optical power required to achieve a specified level of performance, such as a BER.
Shot noise Noise caused by random current fluctuations arising from the discrete nature of electrons.
Signal-to-noise ratio The ratio of signal power to noise power.
Simplex cable A term sometimes used for a single-fiber cable.
Simplex transmission Transmission in one direction only.
Single-mode fiber An optical fiber that supports only one mode of light propagation above the cutoff wavelength.
SNR Signal-to-noise ratio.
Soliton An optical pulse that does not disperse over distance.
Sonet Synchronous optical network, an interational standard for fiber-optic digital telephony.
Source The light emitter, either an LED or laser diode, in a fiber-optic link.
Spectral width A measure of the extent of a spectrum. For a source, the width of wavelengths contained in the output at one half of the wavelength of peak power. Typical spectral widths are 20 to 60 nm for an LED and 2 to 5 nm for a laser diode.
Splice An interconnection method for joining the ends of two optical fibers in a permanent or semipermanent fashion.
Star coupler A fiber-optic coupler in which power at any input port is distributed to all output ports.
Star network A network in which all terminals are connected through a single point, such as a star coupler.
Steady state Equilibrium mode distribution.
Step-index fiber An optical fiber, either multimode or single mode, in which the core refractive index is uniform throughout so that a sharp step in refractive index occurs at the core-to-cladding interface. It usually refers to a multimode fiber.
Strength member That part of a fiber-optic cable composed of Kevlar aramid yarn, steel strands, or fiberglass filaments that increase the tensile strength of the cable.

Tap loss In a fiber-optic coupler, the ratio of power at the tap port to the power at the input port.
Tap port In a coupler in which the splitting ratio between output ports is not equal, the output port containing the lesser power.
TDM Time-division multiplexing.
Tee coupler A three-port optical coupler.
10BASE-F A fiber-optic version of an IEEE 802.3 network.
10BASE-FB That portion of 10BASE-F that defines the requirements for a fiber backbone.
10BASE-FL That portion of 10BASE-F that defines a fiber-optic link between a concentrator and station.

10BASE-FP That portion of 10BASE-F that defines a passive star coupler.

10BASE-T A twisted-pair cable version of an IEEE 802.3 network.

10BASE-2 A thin-coaxial-cable version of an IEEE 802.3 network.

10BASE-5 A thick-coaxial-cable version of an IEEE 802.3 network; very similar to the original Ethernet specification.

Thermal noise Noise resulting from thermally induced random fluctuation in current in the receiver's load resistance.

Throughput loss In a fiber-optic coupler, the ratio of power at the throughput port to the power at the input port.

Throughput port In a coupler in which the splitting ratio between output ports is not equal, the ouput port containing the greater power.

Time-division multiplexing A transmission technique whereby several low-speed channels are multiplexed into a high-speed channel for transmission. Each low-speed channel is allocated a specific position based on time.

Token ring A ring-based network scheme in which a token is used to control access to a network. Used by IEEE 802.5 and FDDI.

Transducer A device for converting energy from one form to another, such as optical energy to electrical energy.

Wavelength The distance between the same two points on adjacent waves; the time required for a wave to complete a single cycle.

Wavelength-division multiplexing A transmission technique by which separate optical channels, distinguished by wavelength, are multiplexed onto an optical fiber for transmission.

WDM Wavelength-division multiplexing.

INDEX

Figures are indicated by *italics*.
Tables are indicated by *tab*.

D

N

O

P

Q

R

U

V

W

X

Z